P9-AFN-324

SPACEFARERS

SPACEFARERS

How Humans Will Settle the Moon, Mars, and Beyond

Christopher Wanjek

Harvard University Press

Cambridge, Massachusetts
London, England
2020

Printed in the United States of America

First printing

Library of Congress Cataloging-in-Publication Data

Names: Wanjek, Christopher, 1967– author.
Title: Spacefarers : how humans will settle the Moon, Mars, and beyond / Christopher Wanjek.
Description: Cambridge, Massachusetts : Harvard University Press, 2020. | Includes index.
Identifiers: LCCN 2019046763 | ISBN 9780674984486 (cloth)
Subjects: LCSH: Astronautics and civilization. | Interplanetary voyages. | Outer space—Exploration.
Classification: LCC CB440 .W36 2020 | DDC 919.904—dc23
LC record available at https://lccn.loc.gov/2019046763

For my daughter Lin, because hers will be the first generation
to see this stuff happen. And for my wife, Suzumi,
a boundless source of insight and inspiration.

Contents

Introduction: Pre-Launch 1

1. Living on Earth 17
2. Checkup before Countdown 62
3. Living in Orbit 97
4. Living on the Moon 148
5. Living on Asteroids 201
6. Living on Mars 219
7. Living in the Inner and Outer Solar System and Beyond 304

Epilogue: Welcome Home 344

Notes *349*

Additional Reading and Listening *371*

Acknowledgments *375*

Illustration Credits *377*

Index *379*

SPACEFARERS

Introduction

Pre-Launch

Living on a planet or moon far beyond the safe confines of Earth seems a cinch when you have a good animator on your team. Parachutes open and engines fire flawlessly, allowing your spacecraft to sweetly kiss the soft alien regolith, pleasantly free from menacing boulders, cliffs, and canyons. A ground shuttle is available to whisk you to the newly established base a few kilometers from the spaceport with efficiency rivaling a Japanese train. There, the scene is abuzz with able workers busily doing their part—digging, probing, pointing, building, transporting—all as happy as whistling dwarfs. You pass alongside massive, golden-lit domes housing a veritable Garden of Eden, where vegetables grow lush and blight-free. You then step lightly over the threshold into the pressurized habitat, with hardly a thought for the months-long treacherous voyage you took through a soup of cosmic radiation in bone-whittling microgravity. And, when you finally make it to your stylish living quarters, you lie back on your bed and think, gee, if only everything worked this well back home.

Most plans for space settlements look this good on paper and in the animated videos. But the devil is in the details. Mars, should that be your destination, is as frigid and lifeless as the Earth's South Pole, only without the luxury of breathable air. Despite claims that we could go to Mars today with current technology, much still needs to be worked out to ensure this wouldn't be a suicide mission. The formidable solar and cosmic radiation exposure that travelers would encounter during the nine-month flight to Mars (as well as during the return) is considered by some scientists to be a showstopper. Landing safely on the red planet remains a dicey proposition; the majority of landers we have sent have failed. Converting carbon dioxide in the Martian atmosphere to oxygen—robotically, in advance of our arrival—and storing it in pressurized tanks for astronauts when they get there is an untested technology. The same goes for extracting water and fuel in situ for the return trip—feasible, but difficult to pull off even on Earth. And growing potatoes? Alas, the Martian "soil" likely contains toxic levels of perchlorate that would need to be removed—again, by a technology still under development.

The Moon, although much closer to Earth than Mars, is no tiptoe through the tulips. The radical temperature fluctuation—from minus to positive 120 degrees Celsius (250 degrees Fahrenheit) between the two-week-long night and day on the lunar surface—makes for a challenging long-term stay. And then there's all that solar and cosmic radiation showering down on the surface. Plans to deal with these challenges include sending a robot to build protective domes for shelter made from the lunar regolith, an unproven technology. And then? Well, as they say, Rome wasn't built in a day.

Thinking big is important. I began writing this book in the early spring, when my small garden plot was weed-free and redolent with the smell of rich, dark-brown dirt, evenly spread in neat squares. With a dozen or so packets of seeds in my hands, a verdant future lay before me. Much like the astronomy animators, I had it all mapped out. The sunchokes will go there, I said, in the back on the north side, because they are the tallest. I'll plant the beans just in front of them so that they can use the sunchokes as a natural trellis. So clever. I'll plant fava beans, too, because they are so tasty yet so expensive in the stores. In the front I'll plant leafy greens in succession, week by week, which will produce a daily salad straight through to the fall. Tomatoes. You gotta have tomatoes. And big winter squash, like one of those massive blue Hubbards that taste great and store so well. Perfect.

Then came an unusually cold April, colder than March, that killed half of what I had planted, followed by an unusually wet May that washed out nearly the rest. And you know what's funny about winter squash? Apparently, there's this little creature called the squash vine borer, *Melittia cucurbitae,* a species of moth that lays its eggs on the base of a well-developed squash vine in early July so that its larvae can feed safely inside the vine and reach maturity as they kill the entire plant before it can bear fruit. They seem to violate a basic tenet of biology by wiping out the very food source that they will need to rely on during the next season. Who knew?

I relay my garden woes partly as catharsis, yes, but mostly as an example of how often things don't go as planned, despite research and preparations. The unexpected occurs reliably, be it unusual weather, such as a month-long dust storm on Mars kicking up during crucial assembly of habitats, or a fly in the ointment, such as an undetected chemical blocking an essential biological reaction from taking place. NASA had its share of the unexpected in the eight-year rush to place humans on the Moon. The Apollo 1 mission ended tragically for astronauts Gus Grissom, Ed White, and Roger Chaffee when a tiny spark in a high-pressure, pure oxygen capsule environment grew instantly into an unescapable fireball—a design flaw that no one had realized was there. NASA administrators later confessed to being lucky that Apollo 11 ever succeeded; Neil Armstrong had to unexpectedly manually steer the *Eagle* lander away from boulders to a smooth landing site with less than thirty seconds of fuel left.

Apollo 13 astronauts never landed on the Moon as intended. An oxygen tank exploded en route, an accident that would have proved fatal if not for the quick thinking and skilled actions of both crew and mission control. Yet another group of Apollo astronauts just missed a large solar flare that could have poisoned them with radiation. Many space enthusiasts are on NASA's case about not getting us to Mars, as if that were the space agency's sole raison d'être. But despite the agency's flaws, which I don't shy away from enumerating in this book, NASA isn't rushing to place humans on Mars because, reasonably enough, it doesn't want anyone dying out there. You can't launch humans to Mars on a tank of hope. We can and should go to Mars only when the risks and costs are minimized, for there is no imperative demanding immediate action. Right now, the voyage is dangerous, and the cost is exorbitant.

The monumental accomplishment of placing humans on the Moon in 1969 set, for some, unrealistic expectations of what would follow. Pick up any popular space book from the 1970s, and you'll conclude that we should have been to Mars a long time ago. Mars was to be the natural, next step after NASA astronaut Alan Shepard played golf on the lunar surface in 1971, cocky as that was. After all, the Americans and Soviets were sending probes to Mercury, Venus, and Mars by this point, and there was a US space shuttle project in the works with planned twice-monthly flights into orbit. Thomas Paine, who led NASA from March 1969 to September 1970, put a date for Mars on the table: a human spaceflight to Mars with a crew of twelve, to depart Earth on November 12, 1981, powered by a nuclear rocket.[1] Meanwhile, physicists, engineers, and members of the US Congress were holding serious conversations about massive, orbiting spheres to be built in the 1980s that could house more than 10,000 people apiece. The orbiting space residents' primary occupation would be harvesting solar energy to beam down to Earth, weaning the world of its oil dependency. We'd be mining asteroids and living on Mars by the 1990s. The moons of Jupiter and Saturn would be explored by humans by the year 2000.

So, a half century after the first moon landing, why aren't we living "out there" or making entry into space safe and affordable? Many factors are at play, which this book will elaborate on. First, some grounding is needed.

The Kennedy You Never Knew

Some place the blame for our lack of a sizeable presence in space on US president Richard Nixon, who requested a slash in NASA's budget, which at its peak during the Johnson administration in 1966 was an extraordinary 4.3 percent of the federal budget. The NASA allotment had fallen to about 1 percent of the federal budget by the time Nixon left office and has continued to fall, to below a half percent in 2019.[2] It was as if we had invested in the railroad and Nixon tore up the tracks. There's some kernel of truth to this. Historians have documented how landing on the Moon was John F. Kennedy's legacy, carried through by Lyndon Johnson, and that Nixon didn't want to extend that legacy. Scouring through Nixon's archival materials related to NASA, John Logsdon, the director of the Space Policy Institute at George Washington University from 1987 to 2008, neatly summarized Nixon's space strategy as (1) demoting NASA from the sacred position it held in the 1960s to just another domestic program required to compete for funds; (2) reining in human spaceflight to low-earth orbit, within 200 miles of Earth's surface; and (3) focusing on a space shuttle program with no specific goal, forgoing development of large rockets capable of sending humans to the Moon and beyond.[3] Logsdon, in his book *After Apollo?: Richard Nixon and the American Space Program,* noted that Nixon was deeply shaken by the near-fatal events of Apollo 13 caused by a ruptured oxygen tank—so much so that he attempted (in vain) to cancel the Apollo 16 and 17 missions ahead of the 1972 presidential election, fearing that tragedy was inevitable and the criticism would doom his reelection bid.[4]

One should not castigate Nixon, though, whose primary concern was fiscal responsibility. Moreover, focusing on near-Earth activities, as opposed to shooting off to Mars, isn't a poor strategy if the goal is to learn how to more efficiently leave and return to Earth. It was Kennedy, actually, who told NASA administrator James Webb in the Oval Office in November 1962, just two months after his famous Moon speech at Rice University calling for a human Moon landing by the end of the decade, "I'm not that interested in space." This statement lays bare how he and other leaders really viewed the Moon missions back then, and it hints at why there are no Moon villages today. We could have been in a race to the center of the Earth, for all Kennedy cared, because apparently the race to the Moon, in his mind, had no other purpose than "to beat [the Soviet Union] and demonstrate that starting behind it, as we did by a couple of years, by God we passed them," as he said to Webb. "I think it would be a helluva thing for us."[5] This helluva revelation by the man we thought was gung ho about the Moon was not made fully public until 2009, part of 260 hours of secret recordings that Kennedy made in the Oval Office and Cabinet Room, unbeknownst to even his aides.

The Soviets felt the same way as Kennedy did. For Soviet leadership, space exploration needed to serve a purpose to justify the enormous cost and danger. The purpose for the Soviets was primarily militaristic: rockets that could carry cargo into space could also carry nuclear warheads across the globe.[6] The space race was an extension of the missile-based nuclear arms race between the Soviets and Americans that began after World War II; and, metaphorically, the space race was about establishing the higher ground. With the United States beating the Soviets to a *human* Moon landing, and with no intention by the United States of setting up a military base on the Moon, the Soviets no longer had any reason to pursue the Moon—nor did the United States, by extension.* No leader during the heyday of the space race—not Kennedy, Johnson, Nixon, Khrushchev, or Brezhnev—cared so much about human space exploration as to advocate for billions of dollars or rubles to support it without some practical return in the form of military might or one-upmanship. None of these men were sending humans into space because they believed it was our destiny; there were more pressing matters to invest in, as far as they were concerned. No country can participate indefinitely in a space potlatch with taxpayer money.

Thus, going to the Moon had little to do with going to the Moon. This narrative—that we would go to Mars and beyond after the Moon—was created, in part, by space enthusiasts in the 1960s, swept up by the excitement of the Apollo missions. Naturally, many are disappointed today and view the past fifty years as an utter failure to reach that dream—a dream *they* had, not a dream that was a priority for any leader of the 1960s or 1970s, or even for most of the general public. For many US leaders then, space was interesting, to be sure; but the space race was too expensive to be sustainable or to justify to the American public after the 1969

*A military base on the Moon would serve no strategic purpose for Earth-based warfare for any nation because it would be too far away, too high of a higher ground.

Moon landing was accomplished, particularly with the growing expense of the US conflict in Vietnam. The concept of turning the 1970s into a decade-long Apollo-like march to Mars ended before it began, with Nixon sending clear signals to NASA that the White House wouldn't support it.[7]

In retrospect, it may seem to have been foolish to squander an investment in rocketry, namely the Saturn V, arguably the greatest feat of engineering in modern times. But during the 1960s, three deaths on the launchpad of Apollo 1 laid bare the danger; three near-deaths aboard Apollo 13 had many people questioning this danger and expense in the absence of an apparent purpose, since Americans had already beaten the Soviets to the Moon the year before. Even during the peak excitement in 1969, the majority of Americans did not think their country should be spending so much money on space activities.[8] The Moon and other deep-space destinations simply offered no military importance or economic potential around 1972, when the Apollo program was canceled, to justify the continued danger and expense of human exploration. The tax-paying public and the politicians representing them were coming to this understanding. Even most scientists preferred robotic lunar exploration over sending humans to the Moon.

Post-Apollo Missteps

For many a space enthusiast, it nevertheless remains incomprehensible that, a half century after landing on the Moon, all we have today in terms of human space exploration is six or seven astronauts in a tin can of a space station a few miles above Earth maintaining ant colonies and doing flips for schoolchildren. No one in 1970 could have predicted such limited human presence in space in the twenty-first century. Yes, access to space is staggeringly expensive, which has limited the commercial investment in human spaceflight. And yes, there was no compelling reason for humans to be in space in the 1970s, other than for the sheer joy of exploration. That said, the International Space Station? Come on. Is that all there is?

Clearly there were missteps after Apollo that curtailed human activity in low-earth orbit—mistakes that we are living with today. Call it either naïveté or pure hubris, but human space exploration turned out to be much more difficult and expensive than we ever anticipated. The unexpected began to reveal itself in ugly, exorbitant ways. The Nixon administration had a desire to create a fleet of vehicles to lift satellites into low-earth orbit at a relatively low cost, in line with the fiscal realities of the time. "We must . . . realize that space expenditures must take their proper place within a rigorous system of national priorities," Nixon said in March 1970, just a few months after the successful Apollo 11 and 12 missions and a month before the ill-fated Apollo 13. "What we do in space from here on in must become a normal and regular part of our national life and must therefore be planned in conjunction with all of the other undertakings which are also important to us."[9]

Unfortunately, the "normal and regular" Nixon spoke of ended up as an inefficient bureaucracy, a government-funded space program that lacked the necessary directives, fiscal discipline, and managerial leadership to deliver results. The space shuttle program that originated in the early 1970s devolved from the bold promise of a cheap, biweekly, low-orbit ferry into a horrifically expensive fleet of rocket-dependent vehicles averaging only about four flights per year, with two of the five shuttles exploding and killing the crew. The shuttle program's greatest flaw was its emphasis on reusability, necessitating a level of maintenance that proved to be costlier and more time-consuming than using expendable rockets. This led to fewer launches, which further lowered the cost efficiency. As the shuttle was to be NASA's primary launch mechanism, many subsequent projects suffered. Satellites designed for the specific dimensions and weight allowances of the shuttle's cargo bay were postponed or canceled. Cost overruns resulted in less funding for research and development for better rocket technologies, creating a negative cycle that led to NASA's access to space growing more and more expensive, not cheaper. Forget Mars. Forget the Moon, for that matter.

The United States continues to pay dearly today for the shuttle program, quite literally, as the nation lost its ability to place humans in space in 2011 with the retirement of the remaining three space shuttles. As a result, the United States must pay Russia $80 million to fly a single US astronaut into space. Similarly, the International Space Station (ISS) that originated in the early 1980s as "Space Station *Freedom*" for an estimated $8 billion ballooned into a $100-billion venture despite its modest size, having only enough room for seven visitors—a far cry from orbiting cities for 10,000 people that many thought could be built for a similar price.[10] The higher cost of the ISS was driven by the higher cost of the shuttle launches, as well as by poor design and management.

What are the chances of NASA taking us to Mars today or tomorrow, given the agency's post-Apollo performance in the realm of human spaceflight? Many politicians who control purse strings have lost patience with funding dreams that turn into fiscal nightmares, projects that grow so expensive that one debates the merits of canceling them before completion to cease the endless cost overruns. Moreover, in the United States, executive leadership changes every four to eight years, and NASA has had to continually change gears to accommodate the opposing whims of each new administration. As a result, a human flight to Mars has always been "two decades away," from 1970 onward. Indeed, as of the publication of this book (2020), NASA has a plan for sending humans to Mars in, you guessed it, two decades.

Given the expense of working or playing in space, coupled with the examples of how NASA managed its last two human-spaceflight programs, what's needed to get and stay in orbit is a sound business plan. And that, finally, is what's emerging. That's what sets this moment apart from 1970, 1980, 1990, and 2000, when routine human spaceflight was more dream than practicality. There are now so many non-NASA players in the human space-flight game that it is difficult to keep up with all the tangible developments. Whereas before we had only advances in animation to make it look easy, today we have commercial investment and actual products.

What It *Really* Takes to Return for Good

The leading spacefaring nations likely could set up a permanent village on the Moon or Mars within ten years from now. Such a feat, though, requires a major effort; a major effort requires a major financial commitment; and a major financial commitment requires a sound reason. So, what's the reason for human space exploration on celestial bodies? We can't do this just because it's neat. Neat isn't a sound reason. Many futurists and space enthusiasts have been reluctant to probe this critical question. Their vision of the future places us on the Moon, Mars, and clear out to the Kuiper Belt, immersed in nifty technology, none necessarily breaking laws of physics. But few ever delve into *why* we would initiate this, *who* would pay for it, and *how.*

If history can be a guide, nations or individuals will part with large sums of cash for big projects for three reasons: praise of deity or royalty, war, or the promise of economic return. The astrophysicist Neil deGrasse Tyson, director of the Hayden Planetarium in New York, introduced this concept in an essay titled "Paths to Discovery."[11]

Praise of deity is what got us the pyramids and cathedrals; similarly, kings built palaces for their own sense of grandeur. Although neither type of project is seen much these days, war remains a familiar investment. The United States has spent more than $4.79 trillion on wars it has initiated in Iraq, Afghanistan, and related insurgencies since 2003, the price tag of at least forty massive missions to Mars, easily enough to establish a permanent space settlement there.[12] Historically, the Great Wall of China was big and expensive, but crucial from a military perspective. Other war-related projects include the Manhattan Project, the US Interstate Highway System (built to accommodate the transport of military equipment, if needed), and the aforementioned Apollo program. These modern military expenditures spurred economic development. But nevertheless, their purpose was militaristic.

Promise of economic return funded the likes of the Panama Canal and the journeys of Columbus, Magellan, and Lewis and Clark. Governments provided money for exploration with the hope of profit. Columbus was funded by the Crown of Castile not to prove that humans can overcome obstacles (that it is "in our DNA") but rather, primarily, to establish a profitable trade route—and to expand Catholicism (praise of deity) and to beat Portugal (military).

Our renewed interest in human activity in space may actually lead to a permanent presence out there because it is being driven in part by "war," but also by the promise of economic return. This is different from the situation in the 1960s, when war was the sole driver. A war might get us to the Moon or Mars; economic sustainability will keep us there.

War and Profits

War, you may ask? That's right, a new space race is upon us. China's clearly articulated ambitions in space—they have their own space stations (plural) and rockets to take humans there—is prompting the United States and other nations to return to the Moon to establish permanent bases by 2030. If China suddenly proposed a Mars-based settlement by 2032, the United States would strive to place its own there by 2031 and would find the money to do so. There is *currently* no political will to spend $100 billion to place four elite people on the surface of Mars for a few months, as there might be to spend that amount on, say, a missile defense system to keep three hundred million US citizens safe. But priorities would change quickly, should China throw down the gauntlet, as Russia did with Sputnik in 1957.

As for economic return, there are *near-certain* profits to be had in low-earth orbit and *possible* profits to be had on the Moon in the form of tourism and resource mining. The extent of these activities and degree of profit will depend on lowering the cost of access to space so that the return on investment becomes more attractive—something the new space race could help with. Investors are hoping for a snowball effect in which cheaper access to space brings more people there, further lowering the price while a space infrastructure grows. The rocket company SpaceX may be among the best known in the NewSpace scene, but dozens of private firms are building smaller and more economical rockets, miniature satellites called nanosats, and various components and services to accommodate the expected increased human activity in space in the hands of industry.*

*Yes, NewSpace is a thing; and the fact that there's a term in camelCase differentiating the emerging private spaceflight industry from Old Space

Admittedly, there is greater uncertainty for profit beyond the Moon. Unlike the situation in 1492, when Queen Isabella had a sense of a potential market to be exploited by better trade routes, Mars has precious little to make it a profitable colony, at least as things now stand. High cost, high risk, and low return on investment does not make for a viable business strategy. But lower the cost and lower the risk, and then there's a *possibility* of settling on Mars and establishing trade, particularly if a war-driven space race between the United States and China paves the way. Wheels are in motion; engines are firing.

All of this governmental and commercial activity—with companies actually making money business-to-business, not solely with government contracts—reveals not just the hope but an expectation that we will very soon create a viable space-based economy. Although humans have not ventured far from Earth in the past fifty years, we nevertheless have learned and accomplished much. We have placed several automated rovers on the surface of Mars and a multitude of satellites in orbit there, and we have greatly expanded our understanding of the Martian environment to a point that we better fathom the difficulties of living there. We have also successfully landed a probe on Titan, a moon of Saturn twenty-five times more distant from Earth than Mars is, a mind-blowing feat.

In short, we are reaping the fruits of labor led by NASA and the Soviet-cum-Russian space agency from the past fifty years. Wealthy individuals already have purchased tickets to be blasted into space, docked to an orbiting hotel, or shot around the Moon, starting in the 2020s.

Governments plan to outsource space transport to private industry to send researchers to live on the Moon for months at a

(governments and their primary contractors) further highlights how much things are different today.

time, as we do now in Antarctica. Private industry plans to follow, tapping lunar resources for profit. Mars will be in the near offing, as the space infrastructure expands to support it.

The Voyage Has Begun

This book explains how this development will unfold—an exploration of the practical motivating factors for settling new worlds and the earnest plans of engineers, scientists, and entrepreneurs striving to make this a reality. I do not wish to instill false hope and spew whimsical nonsense about teleporting, traveling at warp speeds, or living off-Earth in greater luxury than living on Earth. There can be nothing instant or magical in the establishment of space infrastructure. Human space activity will be fraught with challenges, from the economic to the physical and biological. At its core, though, our presence in space will be a natural extension of what we do now, every day, in terms of science, business, and leisure, to the extent that biology and economy can allow.

The voyage begins on Earth. Chapter 1 explores the three most space-like environments on our home planet: Antarctica, where hardy workers soldier through six months of frigid darkness with no new supply deliveries; nuclear submarines, where navy personnel live in self-contained isolation for months at a time; and high desert plateaus, where scientists try to simulate Martian habitats. What lessons have we learned so far? Chapter 2 prepares for our travels with a medical reality checkup, because the lack of gravity and abundance of cosmic radiation doesn't bode well for long-term survival. Can we overcome these challenges? Chapter 3 jumps into low-earth orbit. What plans are afoot to replace the International Space Station, which some see as a marvel of technology and others see as a colossal lost opportunity and waste of money? How will inflatable habitats provide space tourists with weeklong thrills and set the stage for more permanent structures to house space workers and permanent settlers? How will we get into space: with traditional rockets or perhaps space elevators, space hooks, and other clever ideas?

Chapter 4 places us back on the Moon, where scientific bases will surely mimic the settlements on Antarctica and where mining and even tourism could make the venture very profitable. Following the money, Chapter 5 takes us to the next inevitable step into the Solar System, mining asteroids.

Chapter 6 features Mars, the source of endless fascination. Mars may be the first Solar System body that's actually settled by humans with the intent of raising families there, as opposed to the scientific and mining activities on the Moon. The close of the twenty-first century may see humans spread across the local Solar System, from Earth to the Moon and to Mars. By this time, we may begin to venture deeper into the Solar System to the moons of Jupiter and Saturn, which could harbor life and which could sustain small scientific bases. Closer to the Sun, Mercury is surprisingly habitable with advanced technology, while Venus is in some ways the most habitable planet in the Solar System aside from Earth, as long as you live in floatable cities above the clouds there.

Perfecting ways to live on Mercury or the moons of Jupiter and Saturn sets the stage for deep-space travel to the outer planets of Uranus and Neptune; the minor planets beyond, such as Pluto; and the icy rocks of the Kuiper Belt. I discuss all of these concepts in Chapter 7, along with the notion of space arks made of comets or asteroids to take us to other stars—perhaps our destiny hundreds of years from now. The Epilogue brings us back to Earth at a time when colonies are established throughout the Solar System. How will life on the mother planet change as a result?

So, let us now boldly, but prudently, go where no one has gone before.

1

Living on Earth

Earth has a lot going for it. It has water and warmth, two key ingredients for life that don't seem to be coupled on any other planet or moon that we know of. Pluto, for example, has water and lots of it, in the form of ice. But there's no warmth; it is so far away that the Sun appears as a lonely glimmer of light. Venus has warmth, some 800 loving degrees' worth of it. Fahrenheit or Celsius? Maybe it doesn't matter, because there's no water on Venus anymore. It's been boiled into gas and then stripped apart by solar wind. So, as it stands, Earth is the only place in this vast Solar System that has water in a temperature range that allows it to be liquid, or fluid, to enable life. Some moons of Jupiter and Saturn may have oceans beneath solid ice that could harbor life, as might Mars itself; the evidence for liquid water there is strong, but the possibility of it being suitable for life is speculative.

Earth enjoys other charms you perhaps don't think about. There's an atmosphere. Not many other celestial bodies have this, either. The Earth's atmosphere traps in the aforementioned warmth at just the right level and allows it to circulate. That's not the case on the Moon. The Moon can vary in temperature by hundreds of degrees, depending on where the Sun is shining, because there's no atmosphere to trap air, create wind, and circulate the heat. In addition, the Earth's atmosphere blocks gamma rays, X-rays, and most of the Sun's ultraviolet radiation from reaching the Earth's surface—radiation that would cause cellular mutations and make it impossible for life to establish itself on land, let alone flourish. Our atmosphere also provides pressure that keeps liquid water from freely expanding into a gaseous phase. On Mars, if you were exposed to the elements without a pressurized space suit, the water in your blood would "boil" in a few seconds. In this regard, human space settlements could be easier on the distant, chilly moon of Saturn called Titan because it *does* have a thick atmosphere to provide natural-feeling pressure, and all you would need is oxygen and very warm clothes (more on Titan in Chapter 7).

Earth also has a magnetosphere, a massive magnetic field that deflects solar particles and cosmic radiation stretching beyond the Solar System from destroying life. It also prevents the atmosphere from being blown away by solar particles. Moon and Mars? Nope, no magnetosphere. Titan doesn't have one either, but Saturn's magnetosphere extends beyond Titan to compensate.

And Earth has one more thing in just the right amount: gravity. If the International Space Station has taught us anything about living in space (and, frankly, it really hasn't, except for this one thing), it's that zero gravity is horrible for our health. Our bones leach calcium; our muscles atrophy; our eyes eventually stop working as blood vessels weaken and the shape becomes distorted; and much more. Is the gravitational force on the Moon and Mars—about one-sixth and one-third that of Earth's, respectively—enough gravity to keep us healthy?

We have absolutely no idea.

As you can see, Earth fits us like a glove—and I'm not talking about those bulky astronaut-suit gloves that make it nearly impossible to hold a screwdriver, let alone a guitar. No, a perfect glove. Earth was made for humans because this is where we have evolved. Wherever we go in this universe, we need to carry a bit of the Earth along with us in the form of water, warmth, oxygen, radiation protection, gravity, and air pressure—oh, and maybe a guitar.

Why Venture Out?

So, one must ask, why leave Earth to live elsewhere? Sure, just *visiting* the Moon or Mars might be fine. But wouldn't it be crazy, if not unjustifiable, to settle these worlds and raise children there, exposing them to all the risks that come with the lack of earthly protections? It's one thing to set out for an adventure yourself; it's entirely another to homestead an asteroid with a family in tow. This is a legitimate argument against the concept of space settlements. Another argument is, why go into space when we have so many problems here on Earth? How do you justify the enormous expense of making sure a few visitors to the Moon have water when, according to the World Health Organization, more than two billion humans don't have access to clean water right here on our home planet? This seems like an ethical dilemma. And this cannot be overemphasized: governments pay about $7.5 million per day for each astronaut visiting the International Space Station.[1]

But space exploration isn't the cause of misery on Earth, nor is pursuing it tantamount to running away and ignoring our problems. In fact, living out there could help living down here. I firmly believe that space science *is* earth science and that that has been a goal from day one. What we know about pollution and the accumulation of greenhouse gases comes from space-based observations. The space technology behind communication and weather satellites has raised standards of living for everyone, not just the wealthy. The *human* presence in space, as opposed to robots and other machines, is what mostly contributes to the astronomical price tag of space activity—for now.

I don't care much for most counterarguments that spacefaring advocates raise to explain why we *must* go into space. One argument is the perceived population problem: too many people, too few resources. Early in the twenty-first century, the world population passed the seven-billion mark, and by United Nations estimates, we should near twelve billion by 2100.[2] But if we were never able, for whatever reason, to live anywhere other than on Earth, humans would not keep on reproducing and doubling until the whole place imploded. In a worst-case scenario, things might get ugly for a short period—food and water shortages, small-scale wars over resources—but the human species wouldn't cease to exist. The population would naturally level out to match resources. Space cannot be viewed as a means to reduce Earth's population but rather as a place to allow humans to number in the trillions or more.

The likely population scenario, to be played out this century, is that more people worldwide will rise from poverty and have fewer children, slowing global population growth. That's the current trend. Demographers have documented that fertility rates lower to a stable replacement rate of about 2.33 children per woman as life expectancy increases region by region, as women's literacy and education rates rise, as child mortality rates fall, and as technology replaces the necessity of having a large family to grow or gather food.[3] Technology is on the brink of vastly improving the food distribution system, too, replacing fossil fuels with less-polluting renewable resources and reclaiming deserts for housing and agriculture. Earth could handle billions more people if only we were more efficient. The greatest contributor to current population-induced problems of pollution and hunger is inefficiency. The United States throws away 40 percent of its food; and it wastes more than two-thirds of the energy it produces.[4] And that's just one country. There's vast room for improvement.

The bottom line is that space colonization is too impractical to offer a solution to overpopulation. Other solutions will arise long before we have the technology to support billions of humans in space—the kind of number needed to significantly lower Earth's population. Should we desire the lebensraum, though, the asteroid belt has the resources to support one hundred trillion people.[5] Wow. (More on that in Chapter 5).

Another reason proffered by space-travel enthusiasts for the necessity of humans to be a multi-planet species is that some disaster, either human-made or natural, will wipe out life as we know it. This, too, is unlikely to happen anytime soon. No plague, however horrific, has done this before. The Black Death, from the bacterium *Yersinia pestis,* killed upward of half the population of Europe and made a strong showing in China, but other world regions were spared. (And the European Renaissance was inspired, in part, by a shift in worldview brought about by that devastating plague, particularly deadly in Florence, Italy, which lost 60 percent of its population, or some 70,000 people, in just twelve years.[6]) Smallpox virus, introduced by Europeans to the Americas, decimated near-total populations of various indigenous groups there. But still, some survived. Looking to other species, we see that all non-human-influenced and non-celestial-initiated extinctions happened very slowly for them, as a result of evolving into a new species, experiencing intense predation, or losing habitat.

Widespread nuclear war would kill most humans, true. But a handful would survive in well-fortified bunkers or in remote regions near the poles less affected by a nuclear winter. Atmospheric and oceanic scientist Owen Brian Toon estimates that 90 percent of humans would die from starvation after a global nuclear war as the Earth grew too dark and cold to sustain agriculture. Unimaginable horror. Yet that still leaves 750 million people to carry on.* Even a 1-percent-survival scenario leaves millions. Mars would be a refuge from a nuclear holocaust on Earth only if it were already a self-sufficient colony. And true autarky—that is, needing no food or gadgets from the mother planet—is, at best, centuries away. Yes, you have to start sometime. But there's no urgency to start now to build a Martian colony. Indeed, it may be easier to begin to settle Mars only when the technology completely allows for it; that is, starting in 2020 instead of 2050 won't necessarily give us a thirty-year head start over the course of two centuries if 2050 technology allows for the creation of instant settlements via 3-D printing and artificial intelligence at a level we cannot comprehend today. Meanwhile, we can only hope that the nuclear threat will diminish in the generations to come.† If not, that Earth-dependent colony on Mars this century will only watch in horror while Mother Earth is destroyed, as they start counting the days to their own doom, like a bee with no hive.

Asteroids are a persistent menace to terrestrial life. Earth has been hit many times, and every massive impact after life was established here has caused widespread extinction. To be clear, there is a large asteroid out there on a collision course with Earth, bound to strike within the next 100,000 years. But it is most likely that, within a century, before we have self-sustaining space colonies, we will have the technology to detect and deflect an asteroid

*Brian Toon, who coined the term "nuclear winter," explained to me that the 750 million figure is an estimate of how many people might be sustained by primitive agriculture methods.

†Approximately 10,000 nuclear warheads are stockpiled, dramatically down from approximately 70,000 in the 1970s, clearly a positive trend.

threat. An asteroid could sneak in before then, but the chances are extremely low. And if it did, would it spell the end of humanity? The dinosaurs didn't know how to survive; humans do. At this moment, some very wealthy people have bunkers to survive a nuclear winter for years underground. Elected officials and their families, ditto. The anxious and paranoid, too, are stocked up and waiting for Armageddon, however it may come. They could probably last at least a year with the world in flames. Most humans will perish, but quite a few will survive.

Interestingly, a very large "thing" entered the Solar System and passed Earth in 2017. Formally designated 1I / 2017 U1, and nicknamed 'Oumuamua, the 400-meter-long cigar-shaped rock came from interstellar space, reminiscent of the mysterious alien spacecraft in the Arthur C. Clarke 1973 novel *Rendezvous with Rama.*[7] Had that object hit Earth (it wasn't even close), it would have incinerated all life within a hundred-kilometer radius of the impact and caused serious destruction far beyond, but it still wouldn't have caused human extinction.[8]

Another threat, climate change, is real and scary. According to the United Nations, climate change is affecting every country on every continent right now in the form of changing weather patterns, rising sea levels, and increasing extreme weather events, all of which are threatening food security and access to clean water.[9] In a worst-case scenario, the global average temperature on Earth could rise more than 4 degrees Celsius (7.2 degrees Fahrenheit) by the year 2100, which sounds small but is indeed a huge change.[10] Polar ice caps would melt; oceans would rise by many meters; small islands in Micronesia and elsewhere would be swallowed; and most coastal regions would be flooded and largely uninhabitable. Forests would turn arid and burn, and hundreds of millions of refugees would need to seek relief in what are now

An artist's impression of 'Oumuamua, an asteroid-like object and the first confirmed interstellar interloper, originating from an unknown solar system and passing through ours. This unique object was discovered on October 19, 2017. Future human generations may ride local asteroids to other star systems in well-protected cities deep inside the asteroid core.

sparsely populated Arctic and Antarctic regions.[11] This would not be the end of humankind, though. As for our space getaway, here's the catch: you need wealth and well-functioning governments to establish and sustain a space settlement; but in the above scenario, world markets would be in such disarray that no one would be able to afford to leave Earth to start a space settlement. And do you really want to be on Mars, depending on the Earth, when the Earth is in no position to help you?

So, the same logic applies to climate change as it does to nuclear war and asteroids: we need at least a hundred years to have self-sustaining space settlements that could flourish even if the Earth perishes. Yet in a hundred years, if we have the technology to live in space en masse, then we likely would also have technologies to mitigate and even reverse the effects of climate change, such as super-efficient solar energy panels, nuclear fusion, and geoengineering to sequester carbon dioxide. That is, if we have the technology to terraform Mars or the Moon, we'd have the technology to terraform Earth back into Earth. If we could live there—in a terraformed Garden of Eden or a comfortable domed world—then we could live here in the same manner. Having other options in space would be nice, but it wouldn't be a necessity to save humanity from climate change.

One truly realistic and unavoidable threat to life on Earth is rarely discussed, and that's a direct hit from a nearby gamma-ray burst. Gamma-ray bursts are detected nearly every day emanating from distant galaxies, produced by cataclysmic events, such as implosions of massive stars forming a black hole, or the merging of two neutron stars. A burst in the local region of our galaxy, aimed in our direction, within a radius of 7,000 light-years, could instantly deplete most of the Earth's protective ozone layer, cause acid rain, and wipe out many species as a result of rapid Earth cooling and an influx of sterilizing ultraviolet radiation.[12] This may have caused the late Ordovician extinction 440 million years ago (long before the suspected asteroid that killed the dinosaurs), in which 70 percent of marine species perished.[13] You can deflect an asteroid, but you can't stop a gamma-ray burst. Indeed, what alerts you—high-energy photons hitting your space-based detectors—is what kills you a few milliseconds later. Chances of such a hit anytime soon are extremely rare, if that's any comfort. And we could monitor the local universe for any nearby large star about to die.

In 2017, famed theoretical physicist Stephen Hawking said humans will be doomed if we don't get off this planet in one hundred years, a revision of his 2016 proclamation that we had a thousand years to find a new home.[14] He cited war and pestilence. Smart guy, one certainly deserving of the tributes in his name after his death in March 2018. But extinction scenarios that eliminate the human species with such speed require a Hollywood-thriller plot, theoretically possible but not very plausible.

Similarly, the great Carl Sagan wrote, in his book *Pale Blue Dot,* "All civilizations become either spacefaring or extinct." That's not quite accurate, either. Humans will become extinct regardless of what we do. Our evolutionary path could take us toward the smaller-brained, fish-catching aquatic animals envisioned by Kurt Vonnegut in his novel *Galápagos,* in which the author questions the merit of the big human brain. Or, more likely, we will evolve into a more advanced species, subjectively speaking, as did *Homo erectus* and *Homo heidelbergensis* before us. Any human descendant living on planets around other star systems in one million years will no longer be human; we will have speciated long before then.

Thus, although undeniable existential threats to the human species hover over our current incarnation, immediate threats are far too implausible to necessitate or inspire the prompt establishment of space settlements. This is merely stuff of science fiction and doomsday speculation. We are left, really, with no practical urgency to venture into space. Indeed, that's why we aren't in space now, aside from our presence on the International Space Station, floating 250 miles above Earth, 1 / 1000 of the distance to the Moon, if you call that space. (If you picture a globe of the Earth from your days in elementary school, the ISS would be a few millimeters above the surface—the distance from New York to Washington, DC, only up.)

One motivating factor for human space exploration that cannot be discounted, however, is curiosity, the desire to explore. A certain fraction of humanity is enticed by frontiers, driven to adventure. To paraphrase mountaineer George Mallory, of Mount Everest fame, and to move his insight to an entirely different continent, humans initially went to Antarctica for only one reason: because it was there. We went to Antarctica and trekked to the South Pole purely for the challenge and curiosity, not for profit. Only later did a multitude of countries beef up their presence on Antarctica for military reasons in the 1950s, what I'll call the "ice race" that predated the space race by a decade. We'd be there in greater numbers than we are today if it were more profitable. Yet one cannot deny that we went to Antarctica initially out of curiosity, and for the challenge of it.

The origin of Mallory's quotation concerning the reason for attempting to reach the summit of Mount Everest—"Because it's there"—was a *New York Times* article published on March 18, 1923. The reporter followed up with another question about whether previous expeditions had established any monetary or scientific value for reaching the summit, to which Mallory replied, "The first expedition made a geological survey that was very valuable, and both expeditions made observations and collected specimens, both geological and botanical." But, seeing scientific inquiry as a by-product, Mallory went on to say, "Everest is the highest mountain in the world, and no man has reached its summit. Its existence is a challenge. The answer is instinctive, a part, I suppose, of man's desire to conquer the universe."[15]

It's worth noting that Mallory died a year later, attempting to ascend the summit. Another nineteen years would pass before Edmund Hillary and Tenzing Norgay became the first to reach the summit, in 1953. Since then, thousands of climbers also have reached the peak (and more than 200 have died trying). Mallory's spirit inspired them, and I think it will continue to inspire humans who venture into space for no other reason than the fact that it's there.

Mount Everest, the highest mountaintop above sea level on Earth. Reaching its summit has become a symbol of the ultimate human endeavor. But no one lives on Mount Everest. Will the Moon, Mars, and other bodies in the Solar System be merely objects to conquer "because they are there"? Or will people choose to live there?

But venturing is one thing; staying and colonizing is another. We don't live on the summit of Mount Everest. And while we may very well go to Mars to plant a flag and leave, because it is there and the challenge calls to us, we won't live on Mars if there's no reason to keep us there.

A Prelude to Life in Space

As mentioned, to venture into space, we need to take a bit of the Earth along in the form of air, water, food, and various protections. To prepare for the voyage, researchers are attempting to concentrate the hardships of space into earthbound experiments, some natural and some extreme—that is, to bring space down to

Earth—including experiments studying how people perform tasks and interact with each other under the duress of cold, confined, or isolated situations in remote environments such as an Antarctic base.

The discovery and subsequent exploration of Antarctica have remarkably mirrored what we can expect as we venture into space. The icy continent's existence was speculated upon by Aristotle and others more than 2,000 years ago, based solely on the assumed symmetry of the Earth and the notion that there must be lands in the Southern Hemisphere similar to the Northern Hemisphere. Terra Australis Incognita, as it was called, fascinated explorers for the next thousand years. Aristotle got lucky on this one. Land is not distributed symmetrically, but Antarctica was "down" there nonetheless, waiting to be discovered. James Cook got close, spotting islands as his ship dipped below the Antarctic Circle in 1773 and 1774. Credit for the continent's discovery generally goes to Fabian Gottlieb Thaddeus von Bellingshausen, a Russian naval officer of German descent, in 1821 (although Britain's Edward Bransfield and the American Nathaniel Palmer independently may have spotted the continent in 1820).

Early, sea-based exploration soon followed, and the entire continent was largely mapped by the end of the nineteenth century. Then came the Heroic Age of Antarctic Exploration, between 1897 and 1917, when the interior regions were mapped and both the magnetic and geomagnetic poles were reached. We call it "heroic" because many of the explorers perished, including the famed team under the leadership of Robert Falcon Scott, who reached the South Pole thirty-three days after Roald Amundsen in 1911 but died on the return journey. This era concluded with Ernest Shackleton's attempt to make the first transcontinental land crossing, which failed in its objective but succeeded in inspiring the next generation of explorers with Shackleton's team's feats of derring-do and survival.

Expeditions were few and far between for the next forty years before new technology such as air flight and durable equipment allowed for the establishment of permanent scientific bases. The first, more-permanent steps back to exploration were for geopolitical reasons. Germany, which like many other countries had explored Antarctica in the early twentieth century, hoped to set up a whaling station to secure whale oil, used in the production of margarine, lubricants, and glycerin (for making nitroglycerin). The year was 1939, and war was on the horizon. Germany carved out an area they would call New Swabia.[16] The plans didn't go far; no whaling station was established. But the move—as well as Germany's coziness with Argentina, which is close to Antarctica—unnerved the British. The Brits launched Operation Tabarin to establish a large, permanent presence in Antarctica in 1943. This, in turn, set off a bit of a land rush, or ice race, after World War II. Within a decade, nearby Chile and Argentina also set up bases, as did the Soviet Union, Norway, Sweden, France, and the United States.

Things were heating up on the ice. More than a dozen countries had made land claims by 1959 just to have their hand in the game, even though no one knew what that game might be. What was Antarctica good for? This wasn't clear. Decades' worth of surveying revealed minerals and other valuable resources, such as coal and oil. But the harsh climate and remote location in relation to world markets made such resources expensive, dangerous, and thus impractical to extract.

Nonetheless, at the dawn of the Cold War, land meant power, and tensions were beginning to flare. Chile and Argentina, former European colonies, were particularly annoyed by the claims of the domineering Northern Hemisphere countries. The tip of South

America is a mere 1,200 kilometers from Antarctica, roughly five times closer to the continent than are New Zealand or Australia. Perhaps remarkable, given the potential for conflict, twelve countries with significant interests in Antarctica at the time signed the Antarctic Treaty in 1959. The treaty bans military activity on the continent and sets aside Antarctica as a scientific preserve. As of 2015, more than fifty counties have signed the treaty.[17]

You may have noticed that the similarities between Antarctica and the Moon are uncanny: remote, hostile, resource-rich, a natural laboratory, and a source of national pride should one manage to place a flag, base, or settlement there. Some fifty years passed between the first steps on Antarctica and the establishment of permanence. And, lo and behold, fifty years have passed between the first steps on the Moon and plans for going back for good.

Now, it's not clear when the parallels between Antarctica and the Moon first were realized, but the former has certainly provided a template for the latter. Akin to the Antarctic Treaty is the Outer Space Treaty, known formally and prosaically as the Treaty on Principles Governing the Activities of States in the Exploration and Use of Outer Space, Including the Moon and Other Celestial Bodies. The treaty was created during the race to the Moon, when nations were worried about yet another free-for-all (lunar) land grab and, worse, military bases on the Moon. Article II of the treaty states, "Outer space, including the moon and other celestial bodies, is not subject to national appropriation by claim of sovereignty, by means of use or occupation, or by any other means."[18] In essence, the Outer Space Treaty sets the stage for the Moon to be a vast science laboratory, just like Antarctica. (Later I'll discuss how some people view this treaty as thwarting efforts to commercialize space.) There is every reason to believe that working on the Moon, at least at first, will be exactly like working in Antarctica. There will even be tourism. This is why Antarctica is such a great test-bed for Moon and space explorations. Let's examine this more closely.

Life on the Ice

Antarctica is the last great land mass on Earth that has yet to be colonized. Yes, people live there; the population fluctuates from about 1,000 in the nearly sunless winter to about 4,000 in the summer, with a peak between December and February. Yet there are no permanent residents. Some folks stay for a year or two, helping maintain the science bases through the six months of near or total darkness. Then they return home. In decades past, several nations set up fishing settlements. These, too, had no continuation of habitation. Antarctica may be "colonized" only in that it has "a body of people living in a new territory but retaining ties with the parent state," as per the *Merriam-Webster* definition. But no one raises a family there. This, to me—the family notion—is a more complete definition of a colony, as opposed to a scientific outpost or work settlement. To colonize, in my book (quite literally), is to establish communities where adults will live, work, and raise families.

It is perhaps a matter of trivia that I relay how Argentina and Chile laid claims to colonizing Antarctica. As a way to ensure their right to the land, both countries established a civilian base on the continent, the only two among non-civilian scientific outposts. The Chileans refer to their base, Villa Las Estrellas, as a town. Chileans might overwinter there, maintaining the nearby non-civilian science base, but they don't stay for more than a few years. Argentina placed five families at its Esperanza Base in 1977; and, in 1978, Emilio Palma became the first person to be born on the continent there. This "colony"—more stunt or pipe dream than practicality—soon disbanded. There also are eight churches in Antarctica: four Catholic, three Eastern Orthodox, and one nondenominational Christian. Yet the priests, like other workers, only stay for a year or two.

Who goes to Antarctica? The same types of people who will go to the Moon: scientists, engineers, and hardscrabble workers looking for adventure or escape, followed by wealthy tourists. Some will go for a few months, and some will maintain the base year-round for a few years. Approximately thirty countries occupy seventy science bases on Antarctica, of which forty-five are maintained throughout the year. By far, the largest is the US-led McMurdo Station, with a summer population of about 1,200 people and a winter population of about 250. Science endeavors on the continent range from astronomy to zoology. Some experiments performed on Antarctica cannot be done as well anywhere else on Earth, including the IceCube Neutrino Observatory, which observes nearly massless elementary particles called neutrinos as they pass through a kilometer of pure ice at the South Pole. Also, we can study Earth's climate to nearly a million years ago by analyzing carbon dioxide and other molecules trapped deep in the ice. Among the most intriguing sites is Lake Vostok, a liquid lake that lies beneath four kilometers of ice in the coldest region on Earth. Scientists have attempted to tap the lake, and they have found evidence of life in the water sample, although questions remain whether the sample was contaminated with bacteria from the drill itself. Lake Vostok has been sealed off below the ice for millions of years, perhaps like the ice-covered ocean of Jupiter's moon Europa and Saturn's moon Enceladus. Life down there would raise the possibility of life out there.

Antarctica is also the easiest place to find meteorites from Mars, because it's all white and the dark Mars rocks are easy to spot. One, called Allan Hills 84001, contains structures that look like microscopic alien fossils, although most scientists say the evidence is not convincing.

In April 2014, the Scientific Committee on Antarctic Research convened dozens of scientists and policy makers from twenty-two countries to set priorities for Antarctic research for the coming decades. The group settled on six priorities, including climatology and astronomy.[19] A natural by-product of all this science on the ice has been an understanding of how to live and work in an extreme environment, knowledge that will be applied directly to living on the Moon and beyond. Consider life for the fifty hardy souls who staff the Amundsen–Scott South Pole Station during winter. The Sun goes down around March 22 and doesn't rise again till September 21. During that time, no one comes or goes. No planes, and thus no supplies, can be flown in from mid-February to late October; the weather is simply far too harsh to permit flights. Air temperatures drop to nearly –73 degrees Celsius (–100 degrees Fahrenheit). Gale-force winds pummel the station, which is elevated to escape snowdrifts. Working outside, particularly in the winter darkness, requires gear so thick and cumbersome that you might as well be in a space suit.

Life inside the South Pole facility can be cozy but a bit monotonous during the long winter months. The Amundsen–Scott South Pole Station has a relatively large overwinter crew, and that helps keep things lively. Vostok, run by the Russians, is down to thirteen men in winter. Norway's Troll research station has only a six-person skeleton crew. These are Mars-mission-type numbers, and much can be learned about maintaining civility and productivity in such haunting isolation. The Internet is a blessing, but connections can be slow and unreliable.

One element that keeps spirits high is food, especially fresh food. In my 2005 book for the International Labour Organization about workplace meal programs, I described the food service at McMurdo Station.[20] As at other Antarctic bases, fresh food is flown in to McMurdo Station mostly from New Zealand all through the peak summer months, between November and February. Life is good. The food is tasty and free because the management understands that having good meals in remote workplaces is a morale booster, a lesson learned in remote mining operations around the world. Supplying fresh food is quite difficult in the winter, though. For at least seven months, there can be no deliveries. The winter crews must rely on dry, canned, and frozen foods. Such foods tend to lack the crunch or snap of a fresh vegetable, a psychological lift.

Phil Sadler, a machine operator and a jack-of-all-trades, with a background in botany, began experimenting with an indoor hydroponic greenhouse in the 1990s at McMurdo to provide fresh food. This ultimately expanded into a 200-square-meter greenhouse producing upward of 145 kilograms of food per month in the dead of winter—foods such as leafy greens, tomatoes, cucumbers, strawberries, and melons, all grown inside under LED lights. The greenhouse was active until 2013. That's when the National Science Foundation established austral winter flights into McMurdo, making the need for a greenhouse obsolete, as fresh food could be flown in. The Amundsen–Scott South Pole Station still has a greenhouse, properly referred to as a growth chamber because it relies entirely on artificial lighting, not sunlight. This, too, was built by Sadler and then expanded with his colleagues at the University of Arizona Controlled Environment Agriculture Center (CEAC).

The Antarctic Treaty forbids the import of soil, so all plants are grown hydroponically. And anyway, hydroponics is a superior system for growing a variety of greens and fruits in a limited amount of space, provided you have the energy to power the lights.

Sadler's scrappy greenhouse experiments, first made with left-over materials, proved to be so successful in the world's harshest environment that he and other researchers are developing automated systems for other remote Earth locations—yet clearly with space in mind. Sadler and the CEAC have a mature prototype of a lunar-Mars greenhouse to produce 1,000 kcal and 100 percent of one's oxygen needs daily, which I discuss in further detail in Chapter 4. Back on the ice, the Germans have installed a growth chamber about the size of a shipping container, which grows plants under artificial light without soil. This European initiative is led by the Alfred Wegener Institute's Helmholtz Centre for Polar and Marine Research. EDEN ISS was delivered to Germany's Neumayer III Antarctic station in January 2018 as a test-bed for cultivating crop plants in deserts, in areas on Earth with low temperatures, and for future human missions to the Moon and Mars. By August 2018, in the dead of winter, the unit was producing several kilos of tomatoes, cucumbers, kohlrabi, radishes, and other vegetables per week for the ten-member crew there—roughly a salad per person, or about 10 percent of caloric needs. There were hiccups, such as storms knocking out power and damaging elements of the system. But the crew fixed them, as they would have to do on Mars.[21]

Scientific bases on the Moon as early as 2030 will look a lot like those on Antarctica: gutsy researchers performing astronomy, heliophysics, geology, and materials science, setting the stage for commercial mining and eventually tourism. They will minimize their time outdoors and mostly work and live in cramped habitats with food shipped in from Earth, supplemented by vegetables

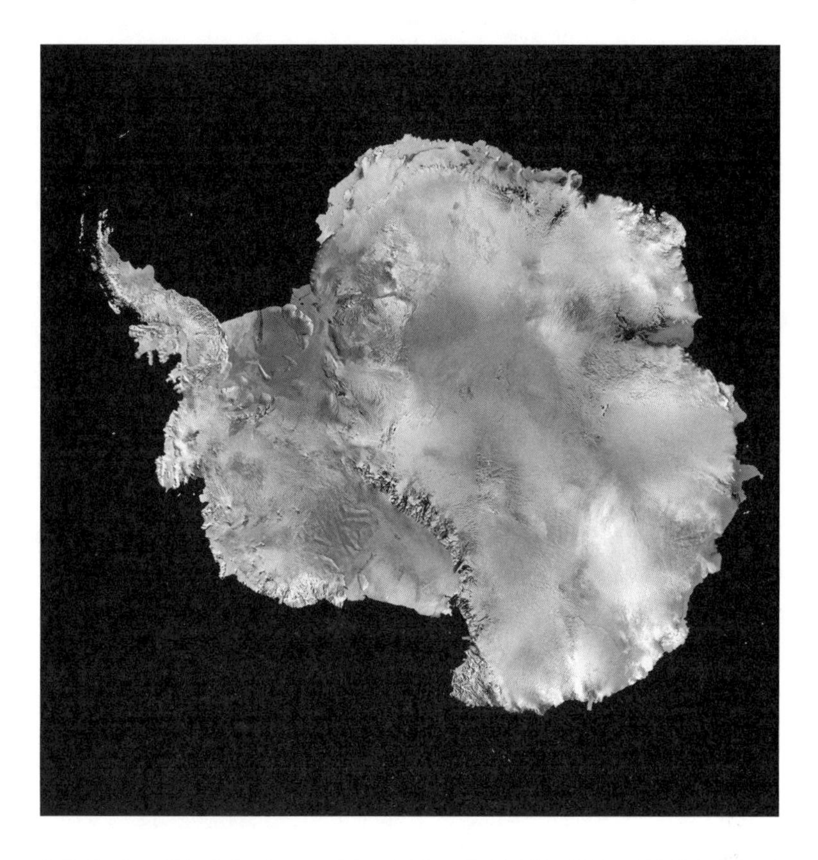

Antarctica as viewed from space. The ice-covered continent is nearly twice as large as Australia but has fewer than 4,000 residents, all temporary. Mars is colder and drier and is lacking the luxury of breathable oxygen and suitable air pressure. Will humans choose to settle on Mars despite the challenges?

grown in artificially lit lunar greenhouses. Water will come from local lunar sources. Workers may spend a few months to a few years there. Coincidentally, one of the best places to establish a lunar base will be at the lunar south pole. (I expand on these concepts in Chapter 4, concerning the Moon.)

Life under the Sea

As hostile an environment as Antarctica is, the icy continent still has one thing that the Moon and Mars doesn't, and that's air. When we go into space, we need to bring along air, or otherwise make it—not only for breathing but also for pressurizing the space suits. That's the added challenge of living off of the Earth.

Yet there is one place on Earth that's exactly like deep space. This is an unworldly environment of complete darkness, low temperatures, abnormal environmental pressure, no natural supply of breathable oxygen, cramped quarters, and chilling isolation, with limited ability to contact any other humans—a place where you and your crewmates need to generate your own electricity to power your lights and machines; to produce air to breath and water to drink; and to maintain the integrity of your food supply. The environment I refer to is deep in the Earth's oceans, on a nuclear submarine.

Life on a nuclear submarine provides the closest example we have of what living would be like on a moon, another planet, or a space voyage. Occupants of a nuclear submarine are in a constant life-and-death struggle with the hostile environment that surrounds them. No hyperbole here. It's a harrowing, nerve-wracking existence. Any number of mishaps—fire, lost pressure, a vessel leak, a gas leak—could spell death for the entire crew. Calamity a mile under the sea is as treacherous as calamity millions of miles in deep space.

Space settlements surely weren't on the minds of the designers of the first nuclear subs, but these vessels contain all the technology needed to live in space. Through their use, we have learned how to live in a completely artificial and self-contained environment. Indeed, these subs are among the most sophisticated engineering masterpieces ever built. At their heart is a compact nuclear engine capable of powering the locomotion of the submarine and all its subsystems for twenty-five years without refueling. The first nuclear sub, the USS *Nautilus,* christened in 1954, could travel 100,000 kilometers on four kilograms of nuclear fuel. That's two and half times around Earth. The USS *Pennsylvania,* one of the largest nuclear submarines in commission, can accelerate its 17,000-ton steel frame to more than 45 kilometers per hour and power the myriad machines and equipment on this 170-meter vessel for years with just a lump of uranium the size of a fist.[22] Typical deployments last three months, often entirely underwater. The ship needs to resurface only to restock food and let the crew see their families.

NASA is eying a nuclear submarine design to explore the ice-covered oceans on Jupiter's moon Europa and the hydrocarbon lakes on Saturn's moon Titan, missions that are many decades away from fruition. And the fuel efficiency alone offered by nuclear fission has space agencies contemplating ways to similarly power bases on the Moon and Mars cheaply and reliably. Harvesting dependable and abundant solar energy on dusty Mars remains questionable, as we will see in Chapter 6. And beyond Mars, using solar energy from a distant Sun becomes largely impractical, leaving nuclear fission as the likely fuel source (unless a nuclear *fusion* engine is developed).

The greater lesson to be learned from a nuclear submarine, however, is not the energy source but rather what one does with that energy: create a self-contained artificial "earth" in an environment not conducive to human life. The first requirement underwater or in space is to generate oxygen. Each member of the 150-person submarine crew needs at least 550 liters of oxygen per day. Without an oxygen generator, the submarine would run out of oxygen in seven days.[23] A perhaps surprising source of this precious oxygen on a nuclear submarine is the ocean water that surrounds it. Each molecule of water, H_2O, contains two hydrogen atoms and one oxygen atom. Through the process of electrolysis, a machine passes an electrical current through distilled seawater to create oxygen gas, O_2, and releases the hydrogen back into the ocean. The process may mimic what needs to be done on the Moon and elsewhere with extraction of the oxygen from water ice deposits.

Oxygen generation, of course, is only part one. We breathe in oxygen but exhale carbon dioxide, CO_2. Submarines needs to remove CO_2 from the air before it builds up to toxic levels. With no plants to take in the CO_2 naturally, a machine "scrubs" the CO_2 from the air by passing the gas through an aqueous solution of monoethanolamine, an organic chemical compound with the formula $HOCH_2CH_2NH_2$. (An aside: not to get bogged down in chemistry, but follow that N in the formula, which stands for nitrogen. The US Navy says the air on a nuclear sub is cleaner than the air you breathe on land, but that's only half the story. Yes, the oxygen is rather pure. But any seasoned submariner will tell you the stench of ammonia-like amines permeates the vessel. That's what the nitrogen converts to in the scrubbing reaction.) We also exhale water vapor, which needs to be removed in this closed system with dehumidifiers. Machines exhale, too. Stoves generate small amounts of carbon monoxide, CO, which is toxic even in tiny amounts. Batteries emit hydrogen gas. Both these gasses need to be filtered, collected, and burned.

For years NASA modeled its air-recycling modules on US Navy designs, but it has since advanced the technology to such an extent that the navy now has come to NASA to help improve submarine air quality, essentially through experimentation with a variety of different kinds of scrubbing actions.[24] Plants help; vegetation absorbs CO_2 and produces O_2. But you'd need hundreds of plants per person in a closed system to reproduce this natural cycle on a submarine or in a space habitat. Water electrolysis is more dependable and requires less energy, given the fact that you'd need to power the lights to grow the food indoors. Plants, at best, complement mechanical air exchange by reducing odors and adding a little oxygen.

Drinking water on submarines also comes from the ocean water, through the energy-intensive process of desalination. Energy also is needed to maintain a constant air pressure of 1 atmosphere as the submarine dives from surface level down to its half-mile-deep cruising depth. This pressure regulation is somewhat the opposite of what is needed on airplanes and what would be needed in space, where celestial bodies contain little or no pressure. The entire weight of Earth's atmosphere presses down on our bodies at sea level with a force of about 15 pounds per square inch (psi). We call this amount, conveniently enough, 1 atmosphere of pressure. On the top of Mount Everest, the atmospheric pressure is only about 5 psi, or a third of an atmosphere, because there is less air on top of you.

On Mars the atmospheric pressure is about 0.09 psi because there's hardly an atmosphere; on the Moon, the atmospheric pressure is essentially zero. But once you're underwater, you have air plus water weighing down on you. The pressure increases about 1 atmosphere for every ten meters of water depth. So, at a half mile, or 800 meters deep, the pressure is up to 80 atmospheres. Outside the safe confines of the sub, you'd be crushed instantly. Submarines maintain constant pressure via a double-hull system comprising an outer waterproof hull and an inner pressure hull made of tough steel or titanium. An advanced ballast system holding various amounts of air or water stabilizes hull compression.

As sophisticated as nuclear submarines are, they are dangerous beasts. Danger lurks not only outside in the chilly, dark depths of the ocean but inside, as well. Most nuclear subs, after all, are war machines armed to the teeth. Fires can easily trigger explosions that can blow the vessel apart. Such was the tragedy for a Russian submarine called *Kursk* on August 12, 2000, when a series of warhead explosions set in motion by a hydrogen peroxide leak and subsequent kerosene-fueled fire tore through the vessel. Most of the 118 crew members were killed in the initial explosions, yet twenty-three members apparently survived for many hours at the far end of the sub until yet one more explosion consumed all remaining oxygen and they suffocated.

Space agencies have learned from both the deaths and lives of submariners. The Russian government's investigation of the *Kursk* disaster, published a few years later in the Russian government daily newspaper of record, *Rossiyskaya Gazeta,* revealed "a shocking level of negligence on all levels of the command; stunning breaches of discipline; and shoddy, obsolete and poorly maintained equipment."[25] That is, the accident didn't have to happen. Is this regard, NASA is militaristic in its approach to workplace safety on the International Space Station, requiring more than an hour each day for routine safety checks. Shortly after the *Kursk* incident, in 2002, NASA officially teamed with the US Navy for the NASA / Navy Benchmarking Exchange, which comprised senior representatives from NASA's Office of Safety and Mission Assurance and the navy's 07Q Submarine Safety and Quality Assurance Division (SUBSAFE). The group identified multiple opportunities for NASA to benefit from SUBSAFE submarine successes.[26] If a deadly accident can happen on Earth, under the sea, it can happen in space.

Submarine living arrangements also are of keen interest to space agencies, because life in space, at least early on, will also be cramped and isolated and thus a potential source of psychological turmoil. Submariners have the added complication of secrecy—silent service, they call it. Nuclear submarines are designed to be stealthy in their surveillance around the globe, so crewmembers cannot call home or have video chats with loved ones or curious schoolchildren, as is the case for astronauts aboard the ISS. Step inside one of these subs on the first day of your mission and hear the hatch close behind, and an instant, regrettable feeling of claustrophobia may come over you. The USS *Pennsylvania* might be the biggest nuclear submarine, at 170 meters long (about two football fields), but it's only 13 meters wide and 12 meters in keel depth. There are no windows, just the haze of artificial lights to guide you through the labyrinth of narrow passageways, a seemingly endless line of metal gadgetry, pipes, and wires from floor to ceiling—raw, like an unfinished construction project. Almost everything is gray, as if set up to be an exercise in depression. There's a constant hum of a machine shop, too, and the odor of lube oil and diesel mixed in with the pervasive amines creates that distinctive *eau de sous-marin*. Low clearance, less than six feet—it's best not to be too tall. Same faces day in and day out. The sleeping arrangements are bunks three levels high, nine to a room that's smaller than a jail cell. No Sun to guide your internal clock; you won't likely see daylight for ninety days.

As witnessed in Antarctica, meals are morale boosters. US submariners describe it as the best food in the navy, by far. Beyond this, the beautiful hypnotic monotony of routines and, perhaps, memories of *Kursk* are what keep the crew going: report to station, check machines, perform maintenance, clean, jump to a surprise drill, train, exercise, jump to another surprise drill, eat, sleep, and repeat. Newbies work on the extensive qualifications required to become a submarine specialist and "earn their dolphins," a uniform breast pin that's a big deal and, in the US Navy, one of the three major enlisted warfare pins. One other element that helps the submariners maintain sanity is their deep sense of purpose that comes with the awesome responsibility of controlling a stealth war machine capable of launching a nuclear attack.

The submariners often lament that they don't have windows. These would be too difficult to install to withstand the high pressure and would serve little purpose, as there is no light below 1,000 meters of ocean. But their lament was the major contributing factor for placing windows on the ISS, a major comfort for astronauts.

The US Navy has studied the psychological well-being of its submariners and has found, not surprisingly, that the cramped conditions lead to poor sleep, irritability, and depression. Conversely, creating the illusion of more space with less-cluttered recreational areas such as mess halls and sleeping berths can improve well-being.[27] Fortunately for the submariners, a deployment rarely exceeds three months. A cramped trip to Mars is expected to last nine months, followed by two years in a cramped habitat, and then a cramped nine-month voyage home. How will the small crew fare? As related below, NASA embarked on a project with the US Navy at the submarine base in Groton, Connecticut, to find out.[28]

Life under the Microscope

Cabin fever. Stir crazy. The more confined and isolated humans are, the more likely they are to develop behavioral conditions or psychiatric disorders. One physician at the Argentine Almirante Brown research station in Antarctica allegedly torched the place in 1984 to avoid spending another winter there, nearly killing himself and his crew.[29] How will this play out on a long voyage to Mars? You surely can envision a nightmarish scenario where *Apollo 13* meets *The Shining*. The fear is real—and, some argue, a showstopper on our quest to travel deep into the Solar System—because this problem could be exacerbated in space, with its lack of gravity affecting sleep and making the crew even more irritable. Astronauts typically don't get more than six hours of solid sleep on any given "night." Both the American and Russian space agencies have noticed telltale signs of cabin fever among those aboard the ISS and Space Station Mir. There have been documented cases of "psychological closing" among astronauts, in which the crew interacts selectively with one or two members of mission control and ignores others as if they were the enemy.[30] Tales of Russian incidents during Mir and Salyut missions are difficult to verify, but anecdotes include cosmonauts turning off the radio communications for days, for spite; disturbing dreams of having toothaches or appendicitis; and impulsive leaps outside the space station without being fully secure.[31] Cosmonauts Valentin Vitalyevich Lebedev and Anatoly Berezovoy reportedly spent most of the 211 days of their Soyuz flight in silence because they couldn't stand each other.[32]

To get an inkling of how a crew would interact with each other on such a voyage and subsequent encampment on a remote world, and to make improvements accordingly, NASA has created artificial environments called analogs to mimic the expected journeys. In short, researchers are studying volunteers as they would lab animals in a cage. One such "cage" is called the Human Exploration Research Analog (HERA), a capsule about the size of a two-bedroom apartment in a nondescript warehouse at the Johnson Space Center in Houston. Four volunteers, mostly strangers to each other, live and work in the HERA habitat / spacecraft for

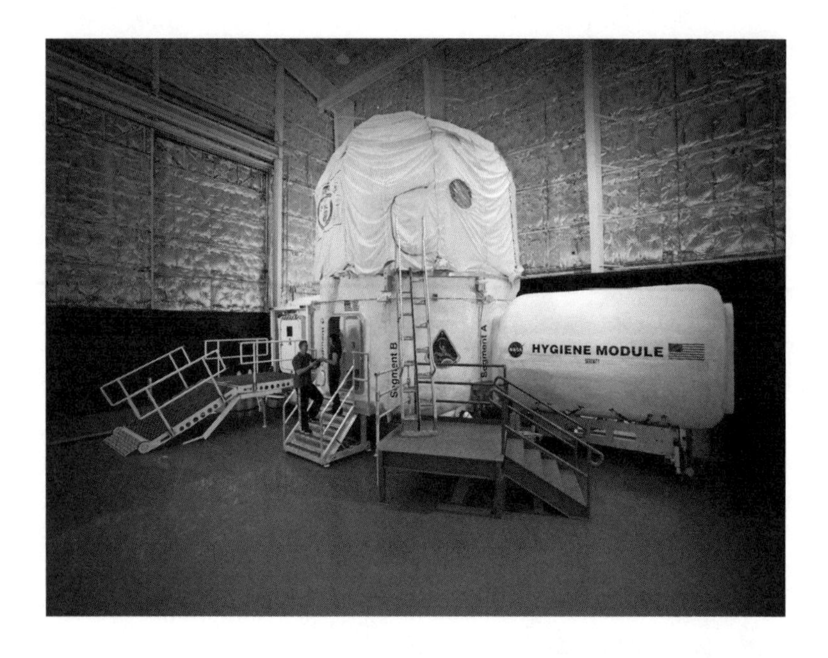

NASA's Human Exploration Research Analog (HERA). Located in a hangar at the Johnson Space Center in Houston, Texas, HERA comprises a central core laboratory segment with an adjoining second and third story devoted to living quarters. Paid participants live entirely within this unit for several months at a time, simulating a journey to Mars or to asteroids.

forty-five days at a time, sometimes longer. They can't leave except to perform a "spacewalk"; there's even an airlock. NASA researchers collect video and audio recordings of the subjects as they go about their "mission" and, in some cases, have sent tapes to the navy submarine lab in Groton to analyze the behavior of the volunteers, tapping into the navy's expertise of tracking submariner behavior.

Sounds a bit like a reality TV show. Yet whereas those kinds of television programs are, ironically, unrealistic, the HERA experiment is crafted to simulate the real journey to Mars as closely as possible in Houston. Any potential participants with "drama issues" are decidedly rejected and not selected for their entertainment value. Potential participants apply from the general population; but, like all astronauts, they must be healthy; must be physically fit, with a body mass index less than 29 and ideally a height less than seventy-four inches; must have vision correctable to 20/20 and no history of sleepwalking; and must possess at least a master's-level degree in engineering, biological science, physical science, or mathematics. With these prime credentials, you can sign up for a forty-five-day stint on HERA for $10/hour to work around the clock on mundane tasks such as virtual maintenance checks, pretend spacewalks, and even piloting the motionless HERA "craft." You exercise on stationary bikes; eat freeze-dried meals; sleep sometimes only five hours a night; and are allowed only short, scheduled, delayed calls to family and friends. NASA will throw a curveball at you, too, and make you navigate through an emergency scenario or two. Nice work for a little more than minimum wage.

NASA conducts several HERA missions per year. They might focus on simulating a landing on an asteroid, or they might take a trip to Mars. The engineering, medical, and team tasks on any of these pseudo-missions mimic what is to be expected on a real mission. And we've learned a lot from HERA already. For example, NASA researchers have found ways to improve sleep and performance by adjusting the lighting throughout the day. Improved habitat designs—same volume of space, different layout—have reduced the sense of claustrophobia.

HERA is just one of more than a dozen analog missions that NASA and its international partners conduct, each one simulating some aspect of space life. There's also NEEMO (NASA Extreme Environment Mission Operations), where "aquanauts" live and work for a month underwater off the coast of Florida to simulate low-gravity environments expected on other planets and moons. They walk about underwater in special suits, collecting soil samples, testing tools and other equipment, and returning to their underwater base called Aquarius, which is about the same size as the living quarters in the ISS, only sixty feet below the waves. The Concordia base in Antarctica—the most remote base in the world, more remote in miles than the ISS—led by the European Space Agency, hosts projects to test the effects of working in frosty isolation where there is no chance of being evacuated during the six-month winter. And the NASA-led bedrest study called VaPER has volunteers spending thirty days lying at a six-degree head-down tilt and breathing air with 0.5% carbon dioxide (ten times more than in regular air), which simulates the high CO_2 environment and excess fluid pressure on eyes and optic nerves that astronauts experience in space habitats. These are the extremes that volunteers and researchers must endure to undo the comfortable earthiness all around us.

One other NASA-funded analog mission that deserves mention here is HI-SEAS, short for the Hawai'i Space Exploration Analog and Simulation. Their focus is living on Mars, and it's HERA on steroids. The HI-SEAS habitat is situated on the barren, arid, high-altitude, Mars-like terrain of the Mauna Loa volcano on the island of Hawaii. A paradise it is not. The base lies next to a string of cinder cones where ash and lava once blew. The escarpment is mostly devoid of vegetation or any signs of life. Its iron-rich soil of crushed basalt closely resembles the Mars regolith in both texture and rusty color. Indeed, NASA exports pulverized lava from these same slopes for tests concerning maneuvering vehicles and planting vegetables in Mars-like conditions. Mauna Loa *is* a Martian mountain, as NASA astrobiologist Chris McKay puts it.

The HI-SEAS dome-shaped unit has about 13,000 cubic feet (370 m^3) of habitable space, roughly equivalent to a volume twenty-four by twenty-four by twenty-four feet, with a loft and a ground floor with a combined usable floor space of about 1,200 square feet. Not exactly roomy. Within the dome lives and works a six-member crew—men and women—acting out a mission on the red planet. They suit up in bulky pressurized suits for their daily ventures outside to dig and conduct science experiments as if on Mars; tend to the solar arrays; spend much of their day performing experiments indoors; eat marginally appetizing food; and so on. Each crew member is assigned a small, private sleeping area in the loft area, and they share a common area, kitchen, toilet, shower, exercise area, laboratory, simulated airlock, and something akin to a mudroom.[33] Any communication with the outside world is delayed by twenty minutes, to simulate the average length of time it would take to send radio waves back and forth between Earth and Mars. As with the HERA experiment, NASA hopes to understand crew dynamics in terms of stress management, problem solving, and morale.

In conjunction with the University of Hawai'i at Mānoa and Cornell University, NASA has conducted multiple HI-SEAS missions. The first, HI-SEAS I, initiated in 2013, comprised a six-member crew isolated for four months and had a special focus on food preparation. Specifically, they compared prepackaged "instant" foods with foods prepared by the crew from shelf-stable, bulk-packaged ingredients. This was to address a so-called gap in knowledge, as defined in the NASA Human Research Program Roadmap. Some members of Congress, however, thought the project was a waste of money. Why did they need to be locked away on a Hawaiian mountaintop for a taste-testing experiment? they asked. No one had a good answer, but we do know that a concoction called "kung fu chicken" was the least favorite pre-prepared meal, according to HI-SEAS I crew commander Angelo Vermeulen.[34]

HI-SEAS II also lasted four months and examined how technical, social, and task roles within the group evolve over time and how they affect performance.[35] This mission established daily routines of food preparation, exercise, and scientific research; conducted geological fieldwork aligned with NASA's planetary exploration expectations; tested equipment; and tracked resource utilization such as food, power, and water. HI-SEAS III upped the stay to eight months, and HI-SEAS IV lasted a full year. The missions continue, each one tweaking living arrangements to reduce stress while introducing new experiments that vitally need perfecting before journeying to Mars. During HI-SEAS II, for example, the crew, with no formal medical training, successfully 3-D printed thermoplastic surgical instruments and completed simulated surgical tasks.[36] During HI-SEAS III, the crew tapped into something called the Virtual Space Station, a suite of interactive, computer-delivered psychological training and treatment programs, meant for confidentially self-medicating nascent stress or depression.[37] Throughout all these missions, NASA staff videotape nearly every movement played out in public areas, taking note particularly of facial expressions that reveal annoyance or disgust.

Stress levels in the HI-SEAS habitat should be lower than on a real mission to Mars because the crew is at low risk of life-threatening danger. No cosmic rays are killing their cells one by one. If their space suit ruptures, there's no fear that the water in their lungs will evaporate within minutes in the extreme low-pressure environment. If they simply cannot take the isolation anymore, or if they have a medical emergency, they could be evacuated in less than an hour by helicopter. Being able to leave in an emergency is a comfort that clearly doesn't exist on Mars, or even a day or so after the launch from Earth, when a U-turn is no longer possible. The HI-SEAS VI mission in 2018 did have a serious incident in which one of the participants was electrocuted only four days in. Medics arrived to evacuate the injured participant, effectively ending the entire mission, not a trivial matter considering the months of preparation to launch HI-SEAS VI.

Those who are really going to Mars can be comforted by the very fact that they are going to Mars. The first to arrive at the red planet will be immortalized. Those stuck on Earth in the HI-SEAS habitat will remain mostly unknown, dedicating a year of their lives under harsh conditions in the name of science and for someone else's future glory. That's stressful in and of itself. Indeed, NASA has observed the "third-quarter phenomenon" among HI-SEAS participants: that's the increased irritability and dwindling motivation and morale that can occur after the halfway point in a long mission, when the novelty has worn off and the reality sinks in that this thing still ain't over.

Cyprien Verseux, the crew astrobiologist for the HI-SEAS IV mission, summed up this sentiment quite nicely, perhaps coincidentally just past his halfway mark. "On Mars, we would know that we are part of history," he said, "whereas here, well, we will be lucky if we have a footnote in a history book." Very true, Monsieur Verseux. So, here's your first footnote—or, perhaps more accurately, an endnote.[38] May you have many more. You deserve them.*

*The accident during HI-SEAS VI may have ended the HI-SEAS "manned" missions. In December 2018, NASA approved a $1 million grant to extend the HI-SEAS program, but only to analyze the data from the first five missions.

From Russia without Love

NASA isn't the only player in town with isolation chambers to study the psychological and physiological effects of long-term confinement. Russia's Institute of Biomedical Problems partnered with the European Space Agency to stage an elaborate project called Mars500 to simulate a human mission to Mars from start to finish. Initiated in 2007, the $15 million project reached full steam in 2010, when an international six-man crew entered a sealed unit for a 520-day mission. The first 250 days were spent in a mock spacecraft taking the men on a virtual journey to Mars. Then, for the next thirty days (they had landed successfully!) they explored Mars, which was a separate, cylindrical module used only for that portion of the experiment. After some basic experiments and flag planting, the men moved back to the first set of modules for the 250-day return to Earth.

I should have prefaced by explaining that the first attempt at this kind of experimental mission didn't go well. The Russians tried it in 1999, but the mission, Sphinx-99, quickly descended into chaos. Four Russians, having been in the simulator for 180 days, were joined by a female health scientist from Canada and two male scientists, from Japan and Austria. The first few weeks were bearable. Then a vodka-fueled New Year's Eve party got wild. Two of the Russian crew members got into a fistfight that splattered blood on the wall; the male Russian commander twice forced himself on the lone female. Morale didn't improve as they sobered up. The Japanese scientist resigned, disgusted by the Russians' unprofessional behavior. The female scientist stuck it out for 110 days but locked the module door between the Russian section and the one she shared with the Austrian. No significant data came from the study.

The Mars500 mission went more smoothly, as Russia decided to exclude females, but still, very little data has emerged nearly a decade since. The project facility was based entirely in a warehouse at the Institute of Biomedical Problems, part of the prestigious Russian Academy of Sciences, in Moscow. That is, the six crew members were detached from the rest of humanity for a year and a half while in the heart of one of the world's largest cities, where eleven million Muscovites went about their days. The surreal facility had five sections, or modules. Three of them—the habitat module, medical module, and storage module—served as the "vehicle" to and from Mars. The modules were reminiscent of a submarine, long and narrow, each between three and four meters wide and twelve to twenty-four meters long. Inexplicably, the modules didn't match any proposed dimensions of a Mars ship and had the comfort of wood paneling and floors. The Mars landing module, at approximately six by six meters, with a triple bunk and limited work space, was far more Spartan and more aligned with what the first landers will be like on Mars. Three crew members lived there during the thirty-day foray to the Mars module. (The lander also was crammed full of food for the second half of the mission and had to be emptied out, another oddity of the project planning.)

More than 6,000 people from forty countries applied for this project in which they would not see blue skies or breathe fresh air for seventeen months.[39] They would need to say good-bye to their loved ones and, worse, be forced to eat Russian food during the full second half of the mission. The pay was alluring, though: $99,000 for the full term. With no place to spend the cash, that sum could go straight into the bank. The project organizers settled on an all-male crew: Romain Charles of France, Diego Urbina of Italy, Yue Wang of China, and Sukhrob Kamolov, Alexey Sitev, and Alexander Smoleevskij of Russia. And they got along rather swimmingly, given the fact that they were on a cheap wood-paneled mission to nowhere. A team of Chinese researchers found that the crew responded more positively to negative situations than expected.[40] Two of the six crewmembers showed neither behavioral disturbances nor reports of serious psychological distress; and only one of the six had signs of serious lethargy and depression, which likely was a result of his insomnia.[41] Otherwise the crew was mildly sleepy, bored, and irritable. No surprises, and no fistfights, yet no reported insights on how to improve morale, as the fake mission didn't seem real aside from its 520-day length.

Grounded in Reality?

Some have asked how practical these analog studies are with paid participants and little threat of serious harm. So, as a complement to the analogs, researchers hope to identify psychological stresses that are probable on a space journey by studying the diaries of the great explorers of centuries prior.[42] Jack Stuster, a cultural anthropologist and contracted NASA principal investigator specializing in ergonomics, sees little difference between being trapped in the ice in a small wooden boat and being in a tin can flying to Mars. In his study of journal entries from those Arctic and Antarctic expeditions, he found elements that led to poor morale that perfectly match those found in modern times on the ISS: lack of communication with the outside world, waste management, personal hygiene, monotony, and more. What held the crews together in times of turmoil was solid leadership with an egalitarian approach, something not common in an era when ships were managed autocratically.[43]

Consider this partial journal entry by Frederick A. Cook, ship physician for the Belgian Antarctic Expedition of 1898–

1899, on May 20, 1898, as the crew was locked in the ice for the winter:

> If we could only get away from each other for a few hours at a time, we might learn to see a new side and take a fresh interest in our comrades; but this is not possible. The truth is, that we are at this moment as tired of each other's company as we are of the cold monotony of the black night and of the unpalatable sameness of our food. Now and then we experience affectionate moody spells and then we try to inspire each other with a sort of superficial effervescence of good cheer, but such moods are short-lived. Physically, mentally, and perhaps morally, then, we are depressed, and from my past experience in the Arctic I know that this depression will increase with the advance of the night, and far into the increasing dawn of next summer.[44]

At least one crew member had died by this point, seemingly from despair. Others were showing signs of paranoia or dementia. One man fell into hysteria and lost his ability to speak or hear. At first, Cook tried remedying their illness with exercise. But walks on the ice around the ship devolved into what was labeled a "madhouse promenade." Cook then devised what he called a "baking treatment" in which the ill sat before the warm glow of the ship's stove for an hour a day. That dedication to the crew's needs lifted their spirits. Cook had speculated that the poor moods were caused in part by a lack of light and vitamins. Exposure to the fire and a diet infused with fresh penguin meat may have reversed that, but Cook ultimately concluded that raising hopes and instilling a "spirit of good humor" was what really got the men through the winter. This strategy was employed by future expeditions.[45]

The upshot is that the journey into deep space will be hard; moods can be lightened with the right design of the spacecraft; and that proper crew selection and leadership may be the most important factors in preventing mutiny or treachery. But, once we land, can we set up self-sufficient habitats with the long-term goal of permanent, protective enclosed cities, just like they do in science fiction books and movies? Judging from one infamous earthbound study, this may be difficult, too.

Life under the Glass

Cruising north out of Tucson, Arizona, on Route 77, the splattering of hastily constructed suburbs and shopping malls gives way to a sunbaked desert brush of palo verde, mesquite, ocotillo, and myriad cacti. Few human-made structures are visible as you pass by Catalina State Park until you approach the sleepy town of Oracle, where, far off to the right of the highway, rises a sprawling complex of pyramids and domes made of steel and glass known as Biosphere 2. The futuristic design, encapsulating more than three acres of desert land converted into a multitude of living biomes—ocean with coral reef, mangrove wetlands, tropical rainforest, savannah grassland, desert, and farm—invokes the ancient Hanging Gardens of Babylon while paying homage to the New Age movement with its focus on holism and a divine Mother Earth. Built to be a hermetically sealed, self-sustaining vivarium of the type we would need to settle on Mars, with no air exchange from the outside, Biosphere 2 was the largest closed system ever created. It is an engineering marvel and, in the 1990s, it was the site of a marvelous failure. But failure is marvelously instructive, as we shall see.

Readers of a certain age may remember the genesis of Biosphere 2 and those eight bright-eyed idealists in their snappy blue jumpsuits saying farewell to Biosphere 1 (that's Earth) to live for two years in what was to be a prototype of a working colony on another planet—all of Earth packed into 3.14 acres, with hummingbirds, monkeys, earthworms, and nearly 4,000 other animal and plant species. The engineering of Biosphere 2 was undisputedly solid, an ecological crystal cathedral. Project founders John Allen and Ed Bass enlisted experts to construct the immense, airtight greenhouses as well as the biomes, or living environments, that would be contained within. For example, the respected Walter Adey, a geologist at the Smithsonian Institution, was in charge of the ocean. Sir Ghillean Prance, a world-renowned botanist and, subsequently, director of the New York Botanical Garden, was in charge of the rain forest.

The steel frame and glass panels of Biosphere 2 were joined more tightly than in any building ever constructed, with less air exchange loss than the International Space Station. The private living quarters were spacious and modern; the kitchen was resplendent with natural light. The base of the structure was stainless steel to prevent any exchange with the soil below. And the hidden subterranean level—an immense array of plumbing, electrical conduits, and air-handlers called the technosphere—circulated the air and water and provided power, with a catacomb ambiance in stark contrast to the lush life above it. The construction was completed in four years, by early 1991, to the tune of $150 million, financed completely by Bass. All the animals, plants, fungi, algae, and bacteria were sealed inside, awaiting the first crew later that year. The senior among them was Roy Walford, sixty-seven, the group's doctor. The others—four women and three men—were twenty- and thirty-somethings with educational backgrounds in either the sciences or engineering. With much fanfare and lavish attention splashed their way by the news media—*Discover* magazine called it "the most exciting scientific project to be undertaken in the U.S. since

President Kennedy launched us toward the moon"—the eight Biospherians, as they were calling themselves, stepped into their new home on September 26, 1991, pledging not to step back out for two years.

Things started to go poorly rather quickly. Twelve days into the mission, one of the Biospherians cut off the tip of her middle finger in a threshing machine while winnowing rice. Walford tried to patch it up, but ultimately she needed to be evacuated for medical attention. And when she returned, she smuggled in some mysterious materials yet to be revealed to help with life under the glass. Hmmm. Also, very early into the mission, the ambient levels of carbon dioxide began to rise, peaking at twenty times that in Earth's atmosphere, and no one knew why at first. Animals started dying. The pollinators—butterflies and bees—were among the first to go. (Later it was a learned that the Biosphere 2 glass limited the amount of polarized light, which the bees needed to find flowers, a crucial lesson for bees on Mars, should we get to that point.) Vegetable production was low, partially as a result of this. The chickens weren't laying that many eggs, either. They and the dairy goats were soon slaughtered for their meat, because they were eating more than they were providing.

Cockroaches, imported purposely to help break down leaf litter, reproduced splendidly, along with the ants, lots and lots of Amazonian ants. No one was exactly sure what happened to those hummingbirds, but the monkeys were the leading culprit, snatching them for food. The monkeys got theirs, though. They died, likely of starvation. Indeed, almost every vertebrate died—fish, bird, mammal. Starvation was setting in for the crew, too. Of the thirty-plus crops planted, only sweet potatoes did well. The crew ate these three times a day, enough to make their hands turn orange with beta-carotene. They also needed to eat their seed crop to survive.

Someone did figure out the O_2–CO_2 problem: seems that bacteria in the vast amounts of muck and compost brought in to the dome to fertilize the farmland were inhaling O_2 and exhaling CO_2 just like humans do, only in far greater quantities. Worse, precious oxygen gas was slowly disappearing. It took months before the crew understood that the concrete in one of the biomes was absorbing the oxygen. After a year in Biosphere 2, oxygen had decreased from about 20 to 13 percent of the atmosphere, resulting in an oxygen level equivalent to what mountaineers breathe at 17,000 feet. The crew was huffing and puffing through the day and suffered from sleep apnea at night. By early 1993, they needed to open the airlocks to allow fresh oxygen to permeate the artificial biosphere, hence breaking the primary protocol of the mission. Meanwhile, the eight-member crew split into two groups of four that didn't speak to each other for the final year of the mission, having argued over the focus of the project. But they all survived, stepping back into Biosphere 1 two years and twenty minutes to the day after they had left, on average about thirty pounds thinner but otherwise healthy.

Mission two began in 1994, with a new set of crewmembers—but, alas, it lasted only six months, as conflicts arose after the hiring of Steve Bannon—yes, that Steve Bannon, later of presidential campaign fame—to reduce project costs. Then came the forced removal of management by armed guards, vandalism, foul language, lawsuits, finger-pointing, shaming—you know, typical Mars colonization kind of stuff. Hey, maybe we should just call off this Garden of Eden project.[46]

Biosphere 2 was included by *Time* magazine on its list, "100 Worst Ideas of the Twentieth Century."[47] The magazine was way wrong, though, because by any reasonable measure, Biosphere 2 was a sensational idea. Failure was more a result of poor management and hubris; it didn't make the idea bad. The project did do several things correctly. For starters, Biosphere 2 succeeded in being a fully sealed environment, something that NASA has since learned from. The infrastructure never failed. The facility recycled all of the spent water and sewage, too. NASA has never achieved a total water recycling rate on the ISS, but total or near-total water recycling will be essential for all space habitats.

Because of Biosphere 2, we now know what *not* to do on Mars and elsewhere. Don't mix your farm area with your living space; keep it in a separate dome. Don't rely on plants to supply all your oxygen, at least not at first. Don't build big; go small and ramp up. Don't think you can master the web of life by importing one animal to control another. Don't underestimate the power of bacteria. And please, for the love of god, don't ever plant morning glories. They will take over your rain forest.

Columbia University took over the management of Biosphere 2 in 1995 with the hope of turning it into a huge research laboratory, but the relationship ended in 2009. In 2011, the University of Arizona took over both the ownership and management and received an additional $20 million from Ed Bass to support research. UA scientists work at the now-unsealed facility studying climate change, water flow dynamics, and energy sustainability. They are conducting one-of-a-kind experiments, such as "watching dirt grow" in the Landscape Evolution Observatory, studying how physical and biological processes interactively control the evolution of landscapes over extremely long periods. Phil Sadler, of Antarctic greenhouse fame, introduced earlier in this chapter, is hoping to create a Mars habitat in one of the Bio-

Biosphere 2 in Oracle, Arizona, originally constructed to demonstrate the viability of a closed ecological system to support human life in outer space. Eight "Biospherians" occupied the habitat for two years in the early 1990s. Although the experiment didn't go well—most of the plants and animals died, and the oxygen supply dropped to a near-deadly level—we learned much from Biosphere 2.

sphere 2 chambers, an ironic use for what was once a bold project to create a mini-Earth.

So, from a psychological and engineering perspective, from what we have learned on Earth, establishing space settlements will be difficult but manageable with emerging technology. But how about human biology? Will space kill us? This question is worth investigating before we take off.

2

Checkup before Countdown

Outer space is the most noxious of substances: devoid of air and filled with a soup of deadly particles in the form of high-energy photons and energetic bits of atomic nuclei. The lack of gravity there affects every element of your being, as even the proteins in your body can't figure out which way is up.

Books and magazine articles about space voyages often compare the adventure to setting off to new lands across treacherous oceans. Our ancestors paddled across the South Pacific in outrigger wooden canoes made by hand with primitive tools. They set off never expecting to return. They spent days, weeks, and months on the open waters, exposed to the elements, with precious little food and water. Many died along the way, but a few made it to their destination and started a new life. No doubt these early migrations tens of thousands of years ago were perilous, but it's not as if each drop of water burns a hole in your DNA; the sea mist doesn't destroy your brain cells; the choppy waves don't cause fluid to build up in your eyes and cause permanent retinal damage. When you finally get to dry land, you can walk. You don't need a team of doctors and engineers to carry you off the boat because your legs are too weak to support you. And, chances are, you will find food and water at your destination when you arrive.

In short, some things can actually live in water, and that which cannot live in water can cross it on driftwood. But space is both sterile and sterilizing. The challenges presented by every journey on Earth that has occurred in the centuries and millennia past, however arduous, pale in comparison to those of a journey in space beyond the Moon. Suggesting otherwise minimizes the sacrifices that the first generation of spacefarers will make. To be clear, space travel is technically feasible today from an engineering standpoint. We placed humans on the Moon fifty years ago, after all. We've sent probes clear out of the Solar System, and we have made soft landings of probes on the surfaces of Venus, Mars, Saturn's moon Titan, the comet 67P / Churyumov–Gerasimenko, and several asteroids. But sending humans beyond the Moon is considered by many doctors to be so dangerous that it is tantamount to homicide.

How Bad Is Bad?

It's illegal for the United States to send a human to Mars. The reason is that the expected radiation exposure for that astronaut—a federal worker—far exceeds the levels permitted for workplace activities by the US Occupational Safety and Health Administration (OSHA). China could do it; Russia could to it. They don't have such pesky regulations. But the United States, legally speaking, cannot. NASA needs to either find a way to reduce the radiation exposure or else change the rules to allow higher exposure. They are working on the former but assuming the latter, or they may never get a Mars crew off the ground.

Yet radiation exposure is only one danger. The NASA Human Research Roadmap has identified thirty-four known health risks and 232 "gaps" in our knowledge of risks. For example, four known health risks are associated with radiation: radiation poisoning from solar flare; brain damage; heart damage; and regular ol' cancer. But among the gaps are questions about hereditary, fertility, and sterility effects from space radiation. So, it's likely that there are more health risks than we realize. Here are the thirty-four known risks of space travel—risks that go beyond basic mechanical dangers, such as the rocket blowing up.

- concern about clinically relevant unpredicted effects of medication
- concern about intervertebral disc damage upon and immediately after reexposure to gravity
- risk of acute (in-flight) and late central nervous system effects from radiation exposure
- risk of acute radiation syndromes due to solar particle events
- risk of adverse cognitive or behavioral conditions and psychiatric disorders
- risk of adverse health and performance effects from celestial dust exposure
- risk of adverse health effects due to host-microorganism interactions
- risk of adverse health events due to altered immune response
- risk of adverse health outcomes and decrements in performance due to in-flight medical conditions
- risk of an incompatible vehicle / habitat design
- risk of bone fracture due to spaceflight-induced changes to bone
- risk of cardiac rhythm problems

- risk of cardiovascular disease and other degenerative tissue effects from radiation exposure and secondary spaceflight stressors
- risk of decompression sickness
- risk of early-onset osteoporosis due to spaceflight
- risk of impaired control of spacecraft / associated systems and decreased mobility due to vestibular / sensorimotor alterations associated with spaceflight
- risk of impaired performance due to reduced muscle mass, strength, and endurance
- risk of inadequate design of human and automated / robotic integration
- risk of inadequate human-computer interaction
- risk of inadequate mission, process, and task design
- risk of inadequate nutrition
- risk of ineffective or toxic medications due to long-term storage
- risk of injury and compromised performance due to extravehicular activity (EVA) operations
- risk of injury from dynamic loads
- risk of orthostatic intolerance during reexposure to gravity
- risk of performance and behavioral health decrements due to inadequate cooperation, coordination, communication, and psychosocial adaptation within a team
- risk of performance decrement and crew illness due to an inadequate food system
- risk of performance decrements and adverse health outcomes resulting from sleep loss, circadian desynchronization, and work overload
- risk of performance errors due to training deficiencies

- risk of radiation carcinogenesis
- risk of reduced crew health and performance due to hypobaric hypoxia
- risk of reduced physical performance capabilities due to reduced aerobic capacity
- risk of renal stone formation
- risk of spaceflight-associated neuro-ocular syndrome

Of these thirty-four risks, three are potential showstoppers: radiation, gravity (or lack thereof), and the need for surgery or a complicated medical procedure. How serious these risks are is more a matter of opinion than of biological fact.

I should also note upfront that it is on this question of health that many, including myself, have developed a love-hate relationship with NASA. One must respect the sincerity with which NASA is studying the human health issue. No other organization has invested as much funding on the topic; NASA is the undisputed leader in this realm, and the world turns to NASA for guidance. We all love NASA. We'd be completely confined to Earth without the US space agency. But NASA isn't infallible. Opinions vary wildly over the direction and efficiency of its health research, as well as its practicality: NASA's overarching emphasis appears to be on the health and protection of astronauts in a microgravity environment, an environment that can and should just be avoided.

The Gravity of the Situation

Let's explore the gravity issue. As stated earlier, we've learned nothing from the International Space Station other than the fact that living in microgravity sucks. A bit flippant, yes. This notion—that gravity is needed for ideal health and that long-term exposure to zero gravity is very harmful—has never been fully tested. Some science fiction writers in the mid-twentieth century speculated that zero gravity would be life-giving: blood would flow more easily; arthritis would be a thing of the past; back pain would be cured for good; and aging itself would slow down. So, bring grandma along for the ride. We had hints from early in the space program that such a rosy scenario wasn't true. Astronauts returned from just a few days of weightlessness feeling weak. But they recovered; and many thought, well, maybe it isn't so bad. Then we spent more time is space. Russians on the Mir space station for months appeared to have some serious, prolonged health issues on their return. The Russians were tight-lipped about the health of their cosmonauts, though, so we never knew for sure. Many of these cosmonauts, championed as heroes, were rarely seen in public after their return. It was the ISS missions that drove home the message: long-term exposure to zero gravity is detrimental to human health on many levels. Kudos to NASA for that.

Before I continue, I should first define some terms. Zero gravity, however visually convenient, can be a misnomer in the context of near-earth activity. The astronauts on the ISS are not living in the absence of gravity. Rather, they are in free-fall, forever falling over the horizon and missing the Earth. The ISS and other satellites are not floating in space because they have escaped the pull of Earth's gravity; they stay up there because of their terrific horizontal speed. The ISS is moving at 17,500 miles per hour. If, somehow, it came to a complete stop, it would fall straight down to Earth, and down would come astronaut, cradle and all. The Earth's gravitational force, in fact, keeps the moving satellites in orbit as a perfectly balanced counterforce, in a downward motion, to the lateral motion set in place during the launch. Without the Earth's gravitational force (if the Earth suddenly, magically, disappeared), the satellites would shoot off in a straight line. Therefore, more accurate terms for describing the lack of sensation of gravity aboard the ISS are microgravity and weightlessness. Yet, even these terms are neither perfect nor synonymous. Astronauts on the ISS have weight, about 90 percent of their weight on Earth, which is only about 200 miles below their feet. They'd be much lighter on the Moon, actually, at just about 16 percent of their weight. Absolute zero gravity is not attainable, because gravity is the force of attraction between any two objects. But in deep space, far from the gravitational tug of any moon, planet, or star, gravity is attenuated to almost zero. I tend to use the terms zero gravity, microgravity, and weightlessness interchangeably in the context of space travel.

Our understanding of gravity's effect on the body has only two data points: one and zero. On Earth, we live with a gravitational force of 1G. On the ISS, astronauts live in 0G. We really don't know about anything in between. Air force pilots might accelerate their jets so quickly that they experience forces of 5G or higher, which sometimes causes them to black out. That's five times the force of normal Earth gravity, which pushes blood out of their brains. But such forces typically last only a few seconds; the pilots aren't living in a hyper-gravity environment. And anyway, we don't care too much about forces greater than 1G because every place we want to go in our Solar System—L2 orbit, the Moon, Mars, and so on—has a gravitational force less than 1G.

What's so special about 1G? This is simply the force we evolved with. Our bones are as thick as they are because of this precise level of gravitational force. Without the pervasive force of gravity all around them, sending constant signals to the cells, bones begin to demineralize and weaken. Muscles, too, expect a certain resistance when contracting. Without the grip of gravity, muscles atrophy and lose their tone. You can exercise in space. Astronauts on the ISS are required to exercise for two hours each day to *minimize* bone loss and *minimize* muscle loss. This works to some degree. But nevertheless, in zero gravity, bones lose density at a rate of more than 1 percent per month, compared to the rate of an elderly person on Earth losing 1 percent per *year*. To visualize how bad that bone loss is, consider the fact that the major obstacle to fully recycling urine into drinking water on the ISS is that the filters get clogged daily with calcium deposits. That calcium is leached from the bones into the urine; this leaching also puts the astronaut at short-term risk for kidney stones and long-term risk for kidney disease.

And for all that muscle exercise on special treadmills, astronauts still find it difficult to walk or even hold a cup after returning from several months in space. Worse for the muscles is the fact that most can't be exercised. Workouts focus on the major skeletal muscles that move the limbs and torso. But there are hundreds of other muscles—cardiac, involuntary, smooth, and other skeletal—that cannot be exercised. Gravity is their workout on Earth, and on the ISS, they aren't getting it. All those tiny muscles in the face and fingers get weaker. Tendons and ligaments also begin to fail in zero gravity. The spine lengthens, and astronauts become one or two inches taller in space, which causes back pain. The Space Medicine Office of the European Astronaut Centre, run by the European Space Agency (ESA), is designing a high-tech, highly tight-fitting "skinsuit" to help astronauts overcome back problems in space. Let's just say the outfit is very, uh, European.

In the body, much more is going on at a cellular level that depends on 1G. Normally, blood pools in the feet because of gravity. Our circulatory system evolved to push blood upward to the brain, a rather important organ. Without gravity, the circulatory system pushes blood upward like a geyser, unharnessed, leaving your head with a pounding feeling. Your heart starts beating faster to pump blood to lower parts of the body. Your body starts thinking there's a fluid surplus, asking, where is all this blood coming from? So your kidneys go into overdrive to remove excess water via urine. But now you are dehydrated, and your blood starts to thicken. This, in turn, triggers the body to stop making red blood cells, and thus you slowly become anemic, sluggish, short of breath, and prone to infection. And so on and so on. It's a holistic medicine nightmare.

The eyes are particularly vulnerable to all this unnatural sloshing of fluids. More than two-thirds of astronauts report having deteriorated eyesight after spending several months in orbit.[1] The fluid pressure flattens the back of the eyeballs, inflames optic nerves, and damages fragile blood vessels. NASA astronaut John Phillips was among the first to report the problem. Gazing out the window, he thought Earth looked blurrier and blurrier with each passing month. NASA tested his sight upon his return and found that his vision had deteriorated from 20/20 to 20/100 after six months in orbit. The implication is that a crew to Mars needs to pack eyeglasses with various prescriptions to help with each phase of their gradual, inevitable, and permanent vision loss. NASA considers the vision issue to be an astronaut's top immediate-term health risk.

Like the eye, the entire brain floats in fluid. A study of thirty-four astronauts for whom MRI images of their brains were captured before and after their missions found microgravity-induced changes that could be permanent: essentially, compression as their brains shifted upward and a narrowing of the brain's central sulcus, a groove in the cortex near the top of the brain that separates the parietal and frontal lobes. These are the parts of the brain that control fine movement and higher executive function; the longer the time spent on the ISS, the worse these brain changes were.[2]

As I mentioned, NASA was hardly focused on the long-term effects of living in space until the late 1990s. The space agency has long been dominated by engineers and physicists. Few medical personnel were even on staff, and fewer still were conducting biomedical research. NASA's primary medical concern was largely limited to the psychology of space travel, as discussed in the Chapter 1. So, in 1997, with the ISS construction ramping up, NASA decided to outsource biomedical research and created the National Space Biomedical Research Institute (NSBRI), a consortium of a dozen university-based research laboratories. The NSBRI jumped right in and studied the health of nearly 300 astronauts who had participated in space missions during the previous ten years. Sure enough, nearly all of them had health problems as a result of their missions, some more serious than others.

What can we do about it? NASA is so new in the health game that it still is conducting more testing than intervention. For example, the Fluid Shifts investigation on the ISS is examining precisely how fluids flow in and around the eye. NASA claims this could help people on Earth who have conditions that increase swelling and pressure. (The US space agency feels the necessity of bringing things down to Earth when it can, justifying the ISS budget.) Also, the Functional Task investigation detects the effects of space on balance and performance, and the Fine Motor Skills investigation examines any changes in the ability to interact with computer-based devices in weightlessness. For now, it's mostly about monitoring. Little is being or can be done to make the microgravity environment more hospitable. True interventions are limited to rigorous exercise, bisphosphonate drugs to slow bone loss, electrolyte packs for fluid loss, and compression cuffs worn on the thighs to keep the blood in the lower extremities.

With zero gravity clearly so bad for your health, a very important question that could be asked is, how much gravity do we need? After sixty-plus years in space, we actually have no idea because, inexplicably, we haven't tested it.

Imagine a graph with an x-axis denoting the level of health from bad to good and the y-axis denoting the level of gravity from 0G to 1G. We have two data points, zero and one. Zero, 0G, is bad for your health, so this data point is at the bottom of the graph, where the x- and y-axes meet. And 1G is good for your health, so this data point is up high on the graph, above the number 1. Now, how do you connect these two points? Is it a linear connection? At 0.5G, are we precisely between the range of bad and good health? Is 0.9G essentially as good as 1G? Is it somehow better? Or, is there a concave curve connecting the two points? Maybe just a little gravity, 0.2G, is fine? Or, conversely, maybe the curve is convex, and 0.5G, 0.75G, and even 0.9G aren't good enough. These questions are important because our Moon and various Jovian and Saturnian moons are about 0.16G; Mars is about 0.38G. Could we live on these places? Again, we just don't know.

No place we hope to settle in space is 0G, so ongoing ISS studies on microgravity and health are nearly useless, in my opinion. All we have learned is that we need to get out of the zero-gravity environment as quickly as possible. There might be a few jobs that entail working in 0G, such as space tourism or construction. So, at best, ISS research may guide us in determining exposure limits to 0G, which likely shouldn't exceed a few months. However, more so than the psychological burden or radiation issues in space travel, gravity makes or breaks the possibility of colonizing the Solar System. If we cannot live and reproduce in 0.38G, the gravitational force on Mars, the game is over for settling on our neighboring planet—unless you're hoping for an unrealistic future of artificial wombs whipping about in a centrifuge, or infinite energy to make ultra-dense matter to add gravity to a planet's or moon's core. Radiation can be blocked, as we explore below; psycholog-

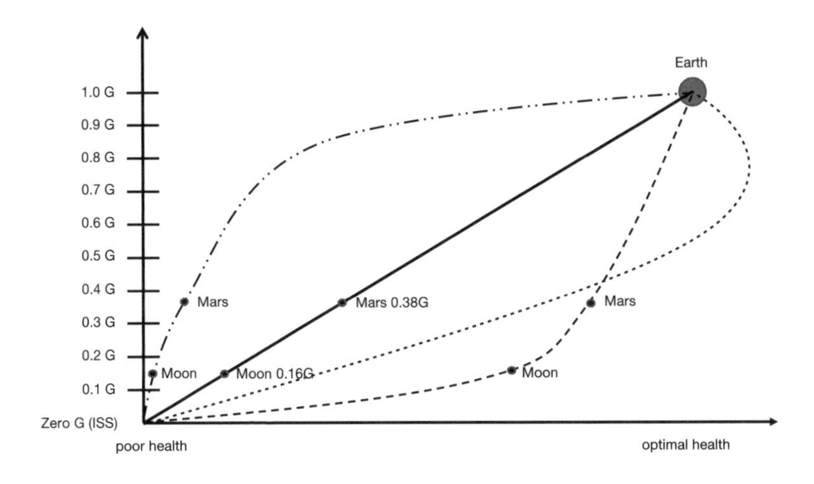

Level of gravity from 0G to 1G versus health. We have two known data points: 0G (the level on the International Space Station) is bad for your health, and 1G (the level on Earth) is good for your health. But how are these two points connected? Is it a linear connection (solid line)? Or is there a concave curve connecting the two points (dashed line)? Maybe just a little gravity, 0.2G, is sufficient. Or maybe the curve is convex (dotted-and-dashed line), and even 0.9G isn't enough for adequate health. There's even the possibility that 0.5G is better for our health, particularly for the elderly (dotted line). Our Moon and various Jovian and Saturnian moons are about 0.16G. Mars is about 0.38G. Could we live on these places? We just don't know.

ical hardships can be overcome. But gravity cannot be increased on a planetary body in any practical fashion.

You would think that the space agencies of the United States, Russia, China, or Europe would have tested the health effects of 0.16G and 0.38G. You can't do this on Earth in any functional, long-term way. There's no antigravity machine; and living underwater isn't the same as reduced gravity. To test lower levels of gravity, we need to create a giant spinning wheel in space, although nothing more complex than the ISS. Centrifugal force, or the force of spin, can simulate gravity. Picture a bucket half filled with water. If you spin that bucket fast enough around and around like a windmill, the water will stay in the bucket even when it's over your head. Slow the rotation, and the water drenches you. But maintain a steady speed and, voilà, artificial gravity keeps the water from spilling.

In space, in zero gravity, the same principle applies. If you spin your habitat around fast enough, the force you feel pinning you to the floor can be equivalent to the sensation of gravity. The math is very clear on this. The perceived force—centrifugal force masquerading as gravity—is a measure of the speed of the rotation and the length of the rotating axis. Put another way, the strength of the force is determined by the angular speed at which it moves and the size of the circular path it is carving out. A small spacecraft needs to rotate quickly, to a dizzying degree. Picture that bucket of water again, or a tilt-a-wheel amusement park ride. You need to spin an object rather fast to create that sensation of being locked in place. But if you have a hollow torus, or doughnut shape, the diameter of football field (about the size of the ISS), then you can produce earthlike gravity with a spin of about four revolutions per minute.

The equation is not complicated: $a=\omega^2 r$, where "a" is the linear acceleration or, in this case, the level of gravity you are after, which on Earth would be an acceleration of 9.8 meters per second squared; "ω" (omega) is the angular velocity, or rotation rate; and "r" is the radius. So, you can see that the radius is inversely proportional to the rotation rate. The beauty of this system is that you can precisely simulate Martian or lunar gravity by adjusting the spin rate of your rotating habitat. You could set it for a year at 0.38G to see if, say, rabbits or catfish could mate successfully. If they could, then you've ensured rapidly reproducing protein sources for Mars. Of course, you could closely examine humans at such a gravitational level to see if we, too, could grow and reproduce. There is debate over how fast a spin humans could handle without getting dizzy, likely no more than four revolutions per minute; the slower the better, though.

So, why haven't we tested artificial gravity? The primary reason is that NASA sold the ISS as a space laboratory for performing microgravity research, not as a stepping-stone to space colonization.* There's some element of charm to a microgravity laboratory. In microgravity, you can perform a few tasks of possible importance to human health, including better crystallizing certain proteins and studying their molecular structures, which could lead to new medicines, although no actual drugs have been developed yet from ISS research. You can also study fluid behavior and materials science in unique ways; but again, no commercial benefit has resulted from these endeavors, either. In assembling the ISS, we indisputably learned how to work in space, crucial lessons for grander construction projects. That's about it, though. Search as you may, you will not find any more benefits from microgravity research. All the benefits for humans on Earth that NASA likes to promote, as useful as they may be, are really spin-off technologies that have nothing to do with microgravity per se: better water and air filtration, portable ultrasound machines, miniaturization, and precision robotic arms now used for surgery.

NASA's original vision for the ISS is clear in an article from its website in 2001 (no longer available on the NASA site, but luckily archived on the Internet Archive Wayback Machine):

*A secondary reason is the expense. To create 0.5G without the dizzying sensation of spinning, the rotating wheel would need to be large, more than 200 meters in diameter at a spin rate of 1 rpm. But it could be smaller and thus cheaper if set for 0.38G at a spin rate of 4 rpm for testing purposes, not lifelong habitation.

> We are not seeking to create artificial gravity in space at this time. NASA and others prefer to work in microgravity—or free-fall—where many amazing experiments and processes can be carried out. The Space Station is the world's only large, long-term, gravity-free science laboratory for developing wonderful new materials, medicines, foods and more. Maybe someday when people spend much longer times in orbit, we'll want to create some gravity with a spinning station (or a spinning section of one) to help keep astronauts' bones strong and deal with other problems of lengthy weightlessness. But not today.[3]

Not "today" in 2001, and not today two decades later. Aligned with this thinking was the designation of the US segment of the ISS as something called the National Laboratory, in 2005. The designation had goals such as advancing STEM education and partnering with the private sector for experiments that require microgravity. NASA has changed its tune in recent years about the intent of the ISS, adopting more of a human health angle, because of the pressure to send people to the Moon or Mars. So lately the spin is that research on the ISS is helping us send astronauts to these destinations—by learning how to combat the ill effects of microgravity on the human body. The ISS costs several billion dollars per year to operate, a hefty sum for protein crystals. In the absence of human research on the partial gravity to be experienced on the Moon or Mars, human research on microgravity is all we have.

One idea is to merely add another module to the ISS that would rotate and produce its own artificial gravity, without interfering with the rest of the ISS. Japan's National Space Development Agency built a 4.5-meter-wide rotating cylinder called the Centrifuge Accommodations Module (CAM) that was capable of providing various levels of artificial gravity to small critters and plants. It was scheduled to be attached to the ISS Harmony module but was canceled in 2004 because of ISS cost overruns. The CAM now sits on display in a parking lot at Tsukuba Space Center about an hour north of Tokyo. Once bitten, twice shy, Japan built and launched a much less ambitious centrifuge called the Multiple Artificial-gravity Research System (MARS) for its own Kibo module of the ISS, safe from NASA's office of cancellations. Mice were exposed to either 0G or 1G for thirty-five days. The mice in the 1G centrifuge maintained the same bone density and muscle weight as mice in the ground control experiment, a proof of concept for rotating habitats.[4] That's one small step for mouse, one giant step for mankind.

The Russian space agency would like to expand on this concept. Russian engineers are designing an inflatable, rotating habitat to attach to Zvezda, the Russian module of the ISS, by 2025. Plans are sketchy at the moment. So, not much to report. US-based Bigelow Aerospace has plans for its own set of inflatable, rotating space habitats, too (more on this in Chapter 3).

Interestingly, all of NASA's original plans for orbiting space habitats called for rotation and artificial gravity. They were never built because they would have been too expensive with the technology of the day, circa 1970. Skylab, NASA's first attempt at an orbiting habitat, had a circular space in which astronauts could run around a ring and experience 0.5G. When building a large facility in space did become feasible, by the 1980s, NASA forwent the artificial gravity route in favor of deliberate microgravity. And today, despite knowledge of how debilitating microgravity is on the body, the US space agency appears to have no plans for changing course.

Although it would add cost to a mission to Mars, many space engineers are advocating for spacecraft concepts that would provide upward of 0.5G, enough to allow the astronauts to step off the craft on the surface of Mars without breaking their legs. One mission vehicle proposed in the early 2000s, called Nautilus-X, would have combined a basic spacecraft with a rapidly spinning centrifuge, where astronauts could sleep or rest in a 0.5G environment, thus reducing their exposure to 0G considerably. But NASA never advanced the concept beyond the initial drawings and proposal. Robert Zubrin, in his aptly named 1996 book *The Case for Mars,* proposes a simple, tethered system that flips over on itself (which I describe in Chapter 6).

The upshot is that prolonged exposure to zero gravity is not something that can be mediated with medication and pressure cuffs and that the only practical solution is to generate artificial gravity. I find it unfathomable that NASA, so preoccupied with safety and health, is hardly considering such a solution.

Solar Radiation

Two types of radiation will plague space travelers: nearly manageable solar radiation and the more menacing cosmic radiation.

Solar radiation is energy from the Sun. Radiation is a broad term that merely means the transmission of energy. Most forms of radiation are harmless, if not life-giving; other forms are life-taking. The Sun emits energy across nearly the entire electromagnetic spectrum, from long-wavelength, low-energy microwaves and radio waves to infrared (providing half the Earth's warmth) to visible light and to shorter-wavelength, high-energy ultraviolet (UV) and X-rays. The Sun also flings off particles in a solar wind, which scientists call radiation because they do carry energy. These particles are protons, electrons, neutrinos, and other subatomic bits of matter.

The Earth bathes in the Sun's radiant energy, and we clearly benefit from it. But we're spoiled. We on Earth are largely protected from the more-deadly, higher-energy components of the

Sun's radiation profile. Very little of the bad stuff—the solar particles and X-rays and the highest-energy forms of UV, which act more like particles than waves—reaches the Earth's surface to harm us. It is either deflected by the Earth's magnetosphere or otherwise blocked by the atmosphere, however contradictory that may sound, given the fact that it's the lower-energy radiation that does reach us. There are no such protections in deep space, beyond your spacesuit and the tin can you are traveling in.

Some folks associate all radiation with danger, but it's mostly the *ionizing* radiation that can be deadly: this radiation is so energetic that it can knock loose an electron from its atom, thus ionizing it. Ionizing an atom makes it less stable and more reactive; in biology, this can manifest in broken chemical bonds and mutations in DNA replication—for the most part, not good. Microwaves, radio waves, and infrared and visible light are not ionizing. Think of them as puffballs: you can throw millions of them at a window, but you'll never break the glass. However, the higher-energy forms of UV and all X-rays are ionizing, as are the particles in the solar wind. Think of these as golf balls. It takes only one to break that window.

With all the sunscreen on the market, you no doubt know about UV radiation and its dangers. Comprising three forms, UV radiation is defined by energy, or wavelength. The least energetic is called UVA and causes skin wrinkles, sunspots, and other types of premature aging. UVA from the Sun easily reaches the Earth's surface, even on cloudy days; it is not ionizing and not particularly deadly. Slightly more energetic is UVB, which crosses the threshold into ionizing. UVB causes sunburn and skin cancer; it is mostly absorbed by the Earth's ozone layer, and even clouds, but clearly much does penetrate. (We do need a little UVB to initiate a chemical reaction in our skin to generate vitamin D.) The most energetic and harmful form of UV is UVC, which luckily is completely blocked by the ozone layer and atmosphere. Welding torches emit the stuff, and you can go blind rather quickly by staring at it without protection. Even more energetic than UVC are X-rays, which easily penetrate soft tissue like skin. The kilometers of atmosphere above us block the Sun's X-rays, too, similar to that lead apron in the dentist's office.

Atomic particles emitted by the Sun, known collectively as solar energetic particles (SEPs), behave just like ionizing radiation and can break chemical bonds and cause cancers and other tissue damage. Most of these deadly particles don't even reach the atmosphere, though. We are shielded from them by our first line of defense, the magnetosphere, a giant magnetic field surrounding us that deflects charged particles, things like positively charged protons (+) and negatively charged electrons (–). Our magnetosphere extends tens of thousands of miles beyond Earth and protects most satellites orbiting Earth, including the ISS. So, this radiation is not a grave concern for visitors to the ISS. I should add that astronauts on the ISS, above the atmosphere but within the magnetosphere, aren't bothered too much by UV and X-rays, either, because their spacesuits and the ISS can provide some protection. But there are limits.

Indeed, everything is manageable until the Sun starts acting up. The Sun frequently emits solar flares, a sudden increase in brightness that brings a higher dose of radiation that can last for hours. A related phenomenon tied to strong solar flares is the coronal mass ejection, or CME, akin to the Sun flinging off globs of material. Both are caused by realignments in the Sun's magnetic field lines, causing snaps and a tremendous release of energy. A solar flare is like a muzzle flash and is predominantly X-ray and UV light. A CME is more like a cannonball propelled forward in a single, preferential direction and comprises mostly particles. Both can overwhelm Earth's defenses, particularly in the far northern and southern regions, where the magnetosphere dips closer to Earth and the ozone layer is thinner. CMEs cause the northern and southern lights—aurora borealis and aurora australis, respectively—as electrons from the solar wind collide with gases in the upper atmosphere of the Earth, exciting the gases and emitting energy in the form of colorful lights. However dazzling it was to the eye, the magnetic disturbance caused by one monstrous CME in 1989 took down the power grid of the entire province of Quebec, Canada. One capacitor after the next tripped and went offline. This wasn't the first CME-triggered blackout, and it won't be the last.

Being an astronaut is dangerous business, and one can ask how much ionizing radiation exposure is acceptable. Astronauts are merely workers, after all, and many workers—miners, radiology technicians, nuclear power plant employees—are in occupations that come with radiation risks. So, let's examine this.

Radiation can be measured in several different ways: by level of radioactivity, or the amount of ionizing radiation released by a material, measured in curie (Ci) or becquerel (Bq); exposure, or the amount of radiation passing by, measured in roentgen (R) or coulomb / kilogram (C / kg); absorbed dose, or the amount of radiation absorbed by a person, measured in radiation-absorbed dose (rad) or gray (Gy); and dose equivalent, which combines the amount of radiation absorbed and the medical effects of that type of radiation, measured in roentgen equivalent man (rem) or sievert (Sv). Although not exact under all circumstances, generally 1 R (exposure) = 1 rad (absorbed dose) = 1 rem or 1,000 millirem (dose equivalent). As a frame of reference, according to the US Nuclear Regulatory Commission, a dental or chest X-ray brings a dose of about 10 millirem. A CT full-body scan is 1,000 millirem. Visiting high-altitude Denver for two days will expose you to a millirem. A transcontinental flight is usually fewer than 5 millirem. The average dose per person per year is about 600 millirem, most of which is unavoidable natural background radiation.

The whole-body dose limit for US workers who deal in radiation, set by the Occupational Safety and Health Administration, is 5,000 millirem, or 5 rem, per year. The limit is 15 rem / year for non-penetrating skin exposure and 75 rem / year for exposure to hands. Doses this high are usually the result of an accident. The workers with the highest average radiation exposure are international airline pilots, dosed with an additional 500 millirem per year. OSHA rules are based partly on increased risks of cancer above baseline, and most doctors think the limits are well within a safe range. For example, a 5-rem exposure will increase your risk of cancer by only about 1 percent. When we talk about living and working in space, these numbers start to sound small. Astronauts that lived on Skylab for a few months received a full-body dose of 17.8 rem; cosmonauts aboard Mir for a year received 21.6 rem.[5] This was all just background radiation, with no blast from the Sun.

Astronauts in the path of a solar flare or CME would, in theory, get a potentially deadly dose of radiation. We've dodged that bullet so far. And there's one comforting note: we get a warning. We know the Sun goes through an approximately eleven-year solar magnetic activity cycle, with highs and lows of solar activity. We know roughly when we can expect more solar storms and bad "solar weather." Moreover, the material flung from the Sun in a CME takes one to three days to reach Earth, enough of a lag time to allow Earth-based mission controls to notify their astronauts to seek special shelter. Solar flares, however, being mostly light, take only eight minutes to reach Earth and might as well be instantaneous. By the time Sun-monitoring satellites relatively close to Earth detect the flare and relay the information to us, at light speed, the X-rays and UV have already arrived. Still, astronauts can run or float to shelter while the flare passes, usually within an hour. Shelter could be simply a more shielded section of the ISS or base.

Once we venture past the protective bubble of the magnetosphere, things can get uglier. On the Moon, on Mars, or during the long journey to Mars, astronauts are like sitting ducks. Again, the ability to receive warnings provides a huge relief. A crew would need some part of the base or craft to serve as a sort of storm cellar, a place with extra protection, and scurry to it once the warning comes. Mind you, protection equals material; material equals mass; and mass equals more fuel and money. Ideally, we'd want bases and crafts with total radiation protection throughout. But in the early days of our journeys into space, we may have to settle for only a small protection chamber. This could be an area coated with thicker metal or even a layer of water, which absorbs solar particle radiation well. On a Mars mission, the ship's pantry would do.

There have been close calls. The now-legendary solar storm of August 1972 occurred between two Apollo missions, a few months after the crew of Apollo 16 left the Moon and a few months before Apollo 17 was to land. Francis Cucinotta, who for many years was NASA's radiation health officer at the Johnson Space Center, estimates that any astronaut on the Moon at that time would have been doused with 400 rem of radiation.[6] The flux would have been more than 45 rem/hour for about a half day with a peak of 241 rem per hour. These levels are rather significant; 450 rem is the LD_{50} or "lethal dose" in which 50 percent of those exposed to that level over a short period would die. You'd need a bone marrow transplant to survive. Anything over 50 rem would likely trigger nausea and vomiting. At 150 rem, you'd likely experience diarrhea, malaise, and loss of appetite. At 300 rem, you'd likely start bleeding internally and lose hair. The LD_{100}, in which no one survives, is 600 rem. The aluminum hull of the lunar lander would have offered some protection to the Apollo astronauts, reducing their possible exposure from 400 to 40 rem, the difference between blood cancer and a bad headache.

Mars is much farther from the Sun than Earth is, but the solar radiation exposure one can expect on the surface of that planet is still deadly because of the lack of a thick atmosphere, and no magnetosphere. Folks on Mars would need day-to-day protection from the ambient solar radiation, as well as a special storm shelter when serious solar flares hit. How often would this be? A large flare of the so-called X-class, the most energetic, occurs about ten times a year. You'd want to monitor the space-weather station and protect yourself against these flares if you were waltzing about on Mars. Astronomers speculate that an extreme solar event about 1,000 years ago baked Mars and, if it were to occur again, would kill or seriously sicken anyone on the planet not taking shelter deep in a lava tube or otherwise underground.[7] A solar event on September 11, 2017, sparked a global aurora at Mars more than twenty-five times brighter than any previously seen and doubled radiation levels on the planet's surface, according to data from NASA's MAVEN orbiter. And this came during what was supposed to be a quiet time in the eleven-year solar cycle.

I noted that the average radiation dose per person per year on Earth is about 600 millirem. On Mars it could be has high as 8,000 millirem, according to data from the Mars Odyssey probe. But that's if you are outside most of the day. On the Moon, we know that the Apollo 14 astronauts received a dose of about 1,150 millirem during their nine-day mission, of which thirty-three hours were spent on the lunar surface. Put another way, a jaunt to the Moon for a week would add twice the natural background radiation you normally receive in a year on Earth. Not ideal, but not deadly. If you lived on the Moon or Mars, you'd likely know the risks and take the daily precautions while remaining alert for solar blasts strong enough to induce radiation sickness. That is, the risk from solar radiation beyond the safety of Earth is real but manageable, perhaps like working as a radiology technician and living in sunny Australia as a faired-skin person who refuses to wear sunscreen.

Cosmic Radiation

Alas, there's no warning and little protection from *cosmic* radiation. These atomic-size bullets will come at you from any direction 24/7, nonstop. Cosmic radiation comes from deep space, beyond our Solar System, comprising largely protons and heavier atomic nuclei moving at nearly light speed, set in motion by distant star explosions. Unlike solar radiation, cosmic radiation wouldn't come in batches so intense that they could sicken or kill you instantly. Rather, cosmic radiation merely slowly eats away at your brain.

On Earth, and on the ISS, we are shielded from most cosmic radiation, also called galactic cosmic rays or high-mass, high-charged (or HZE) particles. Occasionally a few slip in, smashing into the upper atmosphere and causing a cascade of secondary and tertiary particles. What usually happens is that the cosmic rays collide with nitrogen and oxygen, the two most abundant atoms in our atmosphere, and break them open, releasing neutrons, electrons, and more exotic stuff such as muons, pions, alpha particles, and even X-rays. But there's a lot of atmosphere to clear, so the radiation tends to decay or be absorbed before it reaches the surface. In fact, cosmic rays were not detected conclusively until 1912, when Austrian physicist Victor Franz Hess carried electrometers on a high-altitude balloon flight. I previously mentioned that jet pilots and, by extension, flight attendants have an elevated exposure to radiation compared with the general population. Most of this is cosmic radiation.

Apollo astronauts have seen the effects of cosmic radiation—quite literally. Frequently a cosmic particle has zipped through their eye sockets, producing a flash. This has since been named cosmic ray visual phenomena. What's happening at a biological level is not clear. A cosmic ray might be colliding into an optic nerve or perhaps passing through the gel-like vitreous humor, creating a cascade of subatomic particles akin to what happens in the atmosphere. The Apollo astronauts, who traveled beyond the magnetosphere on their way to the Moon, sensed the flashes at a rate of about one every three to seven minutes.[8] The astronauts described the flashes in a variety of ways, which may indicate different physical interactions. The reported shapes of the flashes were spots or dots, stars, streaks or stripes or comets, and blobs or clouds, in order of commonality. Closing your eyes won't help. The astronauts reported that the flashes occurred even when they were trying to sleep.

Of course, the eyes are just a tiny part of the body. The existence of cosmic ray visual phenomena implies that the entire body is being pinged by cosmic radiation around the clock; thousands of rays would pass through you every second. Physicist Eugene Parker of the University of Chicago has said that a third of your DNA would be sliced up by cosmic rays every year you spend in interplanetary space.[9] This is far too much damage to be controlled by the body's own DNA repair mechanisms. We also must remember that we won't be alone in space. We are carrying billions of bacteria, viruses, and fungi, in the form of our microbiome that plays an important role in maintaining health. Microflora in our gut help digest our food, for example. Cosmic radiation could kill or otherwise cause mutations in our microbial passengers, presenting unknown dangers. Only very thick shielding or some kind of mini-magnetosphere around the craft or base (which I explore below) can stop these cosmic rays from passing through your body in space. This has major ramifications not only for spaceflight but also space living. Bases on the Moon, Mars, and nearly anywhere we set up camp beyond our magnetosphere, regardless of how distant they are from the Sun, will be inundated with cosmic rays unless properly shielded. When outside, you would be forced to live with flashes in your eyes, causing untold damage, let alone with the other ramifications of this radiation exposure. Contrary to the silliest of science fiction tropes, cosmic rays don't engender superhuman strength.

The research results from experiments on rodents and space radiation have been ambiguous. Charles Limoli, a professor of radiation oncology at the University of California, Irvine, School of Medicine, led a NASA-sponsored study that exposed laboratory mice to a level of radiation similar to that expected on a six-month one-way trip to Mars. His team found that the radiation caused significant long-term brain damage, including cognitive impairments and dementia, a result of brain inflammation and damage to the rodents' neurons.[10] The mice's brain cells showed a sharp reduction in features called dendrites and spines, like a tree losing its leaves and branches, disrupting the transmission of signals among neurons. The radiation also affected part of the brain that normally suppresses prior unpleasant and stressful associations, a process called fear extinction which, if disabled, can cause anxiety. "This is not positive news for astronauts deployed on a two- to three-year round-trip to Mars," Limoli told me at the time of his 2016 study.[11]

However, as often seen in animal studies, the dose rate in this experiment—bursts between 0.05 and 0.25 Gy / min—was much higher than what would be expected in a human mission to Mars, in which the estimated total dose for a six-month mission is 1 Gy, or 100 rad, evenly dispersed over time. The scientists weren't able to place mice in a habitat with constant exposure to space radiation for six months. Instead, mice were bombarded in great bursts of radiation from a particle accelerator at the NASA Space Radiation Laboratory at Brookhaven National Laboratory and then *observed* for six months. But dose rate matters. Drinking six beers in one hour might get you drunk; drinking six beers in six hours, maybe not. Same exposure, different rate. Better-designed studies would be needed to truly test whether astronauts will be sane or "punch drunk from radiation" when they arrive at Mars.

Other researchers have found that proton radiation causes attention deficits and poor task performance in rats in a simulated space environment[12] and that HZE particles caused an increase in amyloid beta plaque growth associated with Alzheimer's disease.[13] From clinical studies, we know that people who undergo certain kinds of radiation treatment for brain cancer can be cured but have notable declines in their cognitive function. The term is radiation-induced cognitive decline. Upward of half of all patients who receive cranial radiation treatment and who survive their cancer for at least six months will be left with progressive cognitive impairment, particularly in the domains of processing speed (thinking quickly) and memory.[14] But again, this may not translate directly to space: the patients are receiving intense radiation over a period of a few months, whereas in space the radiation exposure on a trip to Mars would be spread out across nearly three years.

Francis Cucinotta, the radiation health officer mentioned earlier, was among the first at NASA to raise an alarm about unsafe cosmic radiation exposure to astronauts, as early as the 1990s. Cucinotta

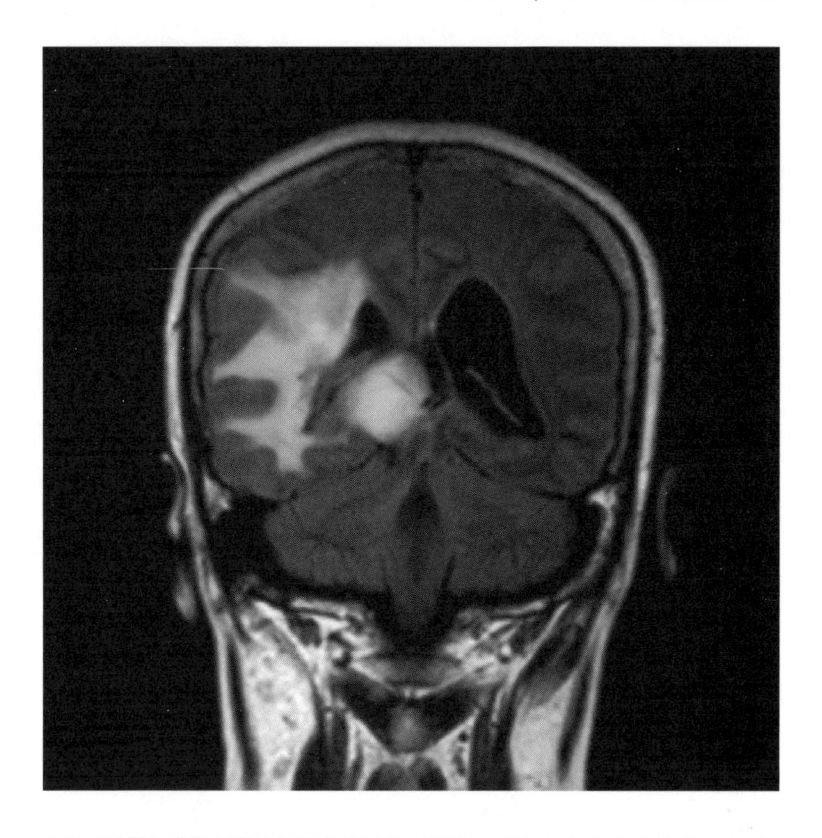

Patients receiving radiation treatment to shrink brain tumors or malformations often experience progressive cognitive impairment as a result of unintentional radiation damage to surrounding brain tissue. Pictured here, a thirty-nine-year-old woman has both cerebral edema (swelling) and atrophy (shrinking) eight years after a gamma knife treatment. Exposure to cosmic radiation on a journey to and from Mars might cause similar damage.

left NASA after more than thirty years there for a position at the University of Nevada, Las Vegas. In 2017 he published a study based on cancer modeling that revealed how cosmic radiation could extend its damage to otherwise healthy nontarget "bystander" cells, effectively doubling cancer risk.[15] As a result of this line of research, and its inconclusive results, NASA has been forced to contemplate the ethics of sending astronauts to Mars. The space agency has proposed an informed-consent approach for accepting probable risk. Some NASA veterans have argued that astronauts are a robust breed who, actually, would be willing to die or have their lifespans shortened for the cause.

The US National Academy of Sciences (NAS), a body of America's leading scientists, took on NASA's request "to develop an ethics framework and to identify principles to guide decision making about health standards for long duration and exploration class missions 'when existing health standards cannot be fully met' or adequate standards cannot be developed based on existing evidence." That committee ultimately ruled, in 2014, "relaxing (or liberalizing) current health standards to allow for specific long duration and exploration missions to be ethically unacceptable."[16] But the NAS committee gave NASA a way out based on the principles of benefit-to-risk and respect for autonomy. Benefit-to-risk is perhaps deliberately ambiguous because there's no necessity to travel into space, and the benefit, or value, is based solely on what we ascribe to it. And respect for autonomy allows the astronauts to be heroes, if they so choose and if there's a benefit outweighing the risk, such as when a firefighter rushes into a burning building to save a child.

If Only a Force Field Existed

What can be done to mitigate the risks? Shielding, lots and lots of shielding. Cosmic radiation is more energetic than solar radiation. Basically, it's moving faster; and some of those atomic bits, such as iron nuclei, are far heavier than the protons and electrons in the solar wind. An iron nucleus would be hundreds of times more energetic than a hydrogen nucleus, which is what a proton is. Flimsy shielding is worse than nothing at all because of the cascade of secondary particles, like shrapnel. A thin layer of spacecraft metal merely scatters the impact of a cosmic ray, turning one fast bullet into scores of only slightly slower bullets. The ship needs thick shielding, and how thick is a simple matter of physics—and economics. (You remember the equation: thickness equals mass, and mass equals money.)

A few centimeters of lead would do the trick. But that would add hundreds of tons to the mission and hence billions of dollars. Water can act as an effective shield. And we need to bring water anyway. So, engineers are playing around with the idea of a hull encompassing the entire craft filled with water. You'd need a lot of it, though, to protect a spacecraft large enough to take a crew to Mars—that is, much more water than you'd need to drink. You can also use trash as extra protection. Still not enough material, but it helps. One very effective shield with low mass would be hydrogen gas, but you'd need a pressurized chamber to hold it, bringing too much mass back into the equation.

The answer to the shielding problem may be a combination of ideas that makes materials serve double-duty. In this regard, hydrogenated boron nitride nanotubes, or hydrogenated BNNTs, show great promise.[17] These tubes are made of carbon, boron, and nitrogen. They are extremely light, hold up to heat and pressure, and are strong enough to serve as load-bearing primary structures for the entire spacecraft. The tubes could be filled with hydrogen gas or water as a primary radiation shield. Boron is an excellent absorber of secondary neutrons, minimizing the radiation cascade effect. As with carbon nanotubes, BNNTs are prohibitively expensive for now but may come down in price in the near future. If the entire craft can't have such a shield, perhaps just the sleeping chambers could, which would effectively reduce radiation exposure by a third if the crew spent eight hours a day sleeping or resting. Perfect protection, as we have on Earth, might not be feasible, but partial protection might reduce the health risks enough to relieve everyone's worries.

Researchers at CERN in Switzerland are working on a magnetic force field to serve as a mini-magnetosphere to naturally deflect the cosmic rays. In 2014, CERN broke a record by creating a current of 20,000 amps at a temperature of 24 Kelvin (about –249°C) in an electrical transmission line comprising a pair of twenty-meter-long cables made of a magnesium diboride (MgB_2) superconductor. While that bodes well for cheaper and more reliable power transmission on Earth, CERN also joined the European Space Radiation Superconducting Shield project to apply the technology to a spacecraft and space habitat. The goal is to create a magnetic field 3,000 times stronger than Earth's own magnetic field, with a ten-meter diameter protecting astronauts within or directly outside a spacecraft. CERN is working on a way to reconfigure the electrical coil for space with MgB_2 superconducting tape.

All of this—the magic materials and the force field—are years away from application. There's no solution to the cosmic radiation problem in the near future aside from the hope that it isn't as bad as the laboratory studies are predicting.

Emergency Surgery

What happens if your appendix bursts en route to Mars? Evacuation is not feasible. The crew will surely include a medic, and hopefully it isn't his or her appendix that is inflamed. Standard procedure for appendicitis on Earth is an appendectomy. Antibiotics work only if the appendix hasn't yet ruptured, and even then, this treatment has limitations. Nevertheless, even a skilled surgeon would have difficulty performing surgery in zero or partial gravity. In the absence of gravity, blood aerosolizes and forms a fog. Tissue density, blood flow, and anesthesia would all be so different as to turn the skilled surgeon into a novice. And should you pull off the surgery, wound healing in space is another unknown. Globally, there are more than eleven million cases of appendicitis annually, resulting in more than 50,000 deaths, mostly entirely from lack of prompt treatment.[18] About half the time, you get little warning. The likelihood of someone on a six-person crew developing appendicitis over the course of a three-year mission is high enough to warrant some folks at NASA and ESA to recommend removing the crewmembers' healthy vermiform appendix before flight, just as a precaution.

The term is prophylactic surgery, and it's not limited to the mostly expendable appendix. Any potential crewmember with wisdom teeth surely would need to have them removed, for fear they might become impacted during a long mission. Some doctors argue for removal of a healthy gall bladder, too, to prevent cholecystitis. Other worries include pancreatitis, diverticulitis, peptic ulcers, and intestinal obstruction, although you can't remove the organs associated with these potentially deadly conditions.[19] You can bet your bottom dollar that you'll be getting a colonoscopy before you leave, though.

Risks are amplified by the fact that we can expect a decrease in immune function during a space journey. Crewmembers will likely experience viral resurgences, such as a herpes relapse. We have a precedent in the Arctic and Antarctic expeditions mentioned earlier. A limb infected from a deep wound or broken bone may need to be amputated, as crude as that might sound in the twenty-first century. Complicating matters is the fact that communication with mission control will not be seamless as crews venture several light-minutes or light-hours into space. Onboard doctors or medics will need to rely on their own wits and some sort of virtual companion, such as a robot or advanced medical software. In a flashback to the pre-medicine era, standard treatment in space maybe be limited to watch and wait.

The Twins Study

Given the minuscule number of humans who have ever had the honor of visiting space—fewer than 700 on this planet of 7 billion—it's nothing short of a mathematical miracle that the astronaut who holds the US record for longest consecutive time spent in orbit has an identical twin brother, also an astronaut, who spent only a short time up there. This coincidence enabled NASA to perform a study on the long-term effects of weightlessness, called the Twins Study. Yes, we've established that weightlessness is horrible for your health. But perhaps there's some reassurance in knowing that you can perhaps recover from a long bout in zero gravity.

The subjects of this serendipitous study were Scott and Mark Kelly. Born in 1964, both were selected to be NASA astronauts in 1996. Mark logged fifty-four days in space spread over four space shuttle missions. He retired in 2011, citing the need to take care of his wife, former US representative Gabrielle Giffords, the victim of an assassination attempt by gun near Tucson in 2011 that left her with a severe brain injury. Scott Kelly, who retired in 2016, spent a total of 520 days in space, including 342 consecutive days starting in November 2012 for the so-called yearlong mission on the International Space Station. The study examined what happened to Scott physiologically and psychologically during that yearlong stretch in space, and compared the data with Mark's, the control subject on Earth. A largely independent, non-NASA team of researchers published their final results in April 2019. Here's what they found.

Most of the biological changes Scott experienced in space reversed to nearly his preflight status. Some changes returned to baseline within hours or days of landing, while a few persisted after six months. Scott's microbiome profile changed radically in space but returned to his preflight profile within a year. By measuring large numbers of metabolites, cytokines, and proteins, researchers learned that Scott's year in space was associated with oxygen deprivation stress, increased inflammation, and dramatic nutrient shifts that affect gene expression. Scott's telomeres, which are endcaps of chromosomes that shorten as one ages, actually became significantly longer in space. The majority of those telomeres shortened within two days of Scott's return to Earth, although no one knows what this means for his long-term health. Researchers also found that, for better or worse, 7 percent of Scott's genes appear to be altered in how they express themselves, via a process called epigenetic change. These genes are related to his immune system, DNA repair, bone formation networks, oxygen deficiency (hypoxia), and excess carbon dioxide (hypercapnia).[20] This much can be expected; it is part of biological adaptation.

So far, few very negative health effects are apparent between the two, aside from Scott now having poorer vision than Mark. However, Scott estimates that he has been exposed to thirty times the radiation of a person on Earth, which will increase his risk of a fatal cancer and earlier death. Yet, he agreed to the yearlong mission on the ISS because he believes it's the only way to understand what to expect on a three-year mission to Mars.[21]

NASA and ESA remain committed to studying the health consequences of unprotected space travel—that is, with prolonged exposure to microgravity and radiation. NASA has a large section dedicated to health called the Human Health and Performance Directorate, renamed in 2012 from the Space Life Sciences Directorate. NASA entered its first memorandum of understanding (MOU) with the US National Institutes of Health, the world's largest funder of biomedical research, in 2007. Among the goals of that MOU was the "development of biomedical research approaches and clinical technologies for use on Earth and in space."[22] Very few of those studies funded through the NIH have actually been performed, however. One study found that "the effect of microgravity on tissue cells per se may cause immunodeficiency."[23] Another found that vitamin K supplements won't do much to prevent bone loss.[24] Other studies have examined what kinds of science could be performed in microgravity, such as DNA sequencing, which proved to be possible albeit lacking any practical application.[25] NASA astronaut Kate Rubins, who performed the DNA sequencing, summed up ISS health studies as better understanding the health insults on bone, muscle, and nerve in order to find ways to more effectively remediate them through exercise or medicines.[26]

In 2017, NASA and NIH signed a new MOU with a more pointed goal of reducing human health risks. One must question the strategy, though, because, short of bioengineering a breed of super humans who can endure microgravity and space radiation, human space migration will be possible only when risks are eliminated with faster, shielded spacecraft that rotate for artificial gravity. We didn't learn to cross oceans by bioengineering gills.

3

Living in Orbit

I will forgo the advice of every writing instructor I have ever encountered and start off this chapter with an onerous mathematical equation: $\Delta v = v_{exh} \ln(M_0 / M_1)$. Uh, that triangle and curly v thing equals what? Actually, this is the Tsiolkovsky rocket equation. It brilliantly details what it takes to get into space and to play there and stay there. The equation also reveals how formidable it is to reach orbit, an inconvenient truth that some engineers call the tyranny of the rocket equation.*

But first we must appreciate the difficulty of getting into orbit: it is not as simple as just shooting something straight up in the air. The necessary energy, speed, and precision are daunting. Orbit implies a lateral velocity around an object. Go too fast, and you leave orbit for a deeper point in space; go too slow, and you can fall back to Earth. The difficulty is that you cannot easily brake

*NASA astronaut and flight engineer Don Pettit appears to have coined this term. He claims, however, he likely will be remembered in centuries to come for inventing a cup to hold fluids in microgravity suitable for clicking together with fellow crew members, overcoming the indignity of drinking from a child's juice bag.

and fine-tune your speed, as you can on land or even in the air. The near vacuum of space offers little resistance; objects in motion stay in motion. Speeding up or slowing down requires energy, burning the precise amount of fuel and thrusting the spacecraft in the precise direction to achieve the desired maneuver. To slow down, you need to fire the engines in reverse. Docking is an even greater challenge. The ISS has an orbital velocity of about 17,150 mph. To dock with the ISS, your craft needs to be traveling at least 17,151 mph to catch up to it. You likely want to go faster. Then, as you approach, you need to slow down, adjusting your speed just right without the luxury of being able to tap on the brakes. You have only one chance to get it right. If the ISS looks nearly still in videos, that's because you are seeing an optical illusion, akin to someone inching up to you at 61 mph on the highway as you drive 60 mph. Drive at 17,150 mph; roll down the window; and try to grab a cup of coffee without spilling a drop from a guy attempting to catch up to you. That's what it's like to be on the ISS awaiting a spacecraft to dock.

What's amazing about Konstantin Tsiolkovsky, the Russian who derived the rocket equation and published it in 1903, is that he figured out orbital maneuvering with the correct precision with paper and pencil as a self-taught amateur scientist. He calculated the velocities needed to place any object in orbit around any Solar System body, not just Earth. Tsiolkovsky had the personality of a dreamer and recluse, absorbed in books, particularly science fiction. He was nearly completely deaf after being stricken by scarlet fever at age ten, and subsequently was denied entry to school. He turned to studying math and physics in order to explore the reality of his dreams while eking out a modest living as a provincial school teacher in what was then a backwoods region on the outskirts of Kaluga, a town 120 miles southwest of Moscow. Tsiolkovsky also conceived the space elevator, discussed later in this chapter, and prototypes of twentieth-century dirigibles. He died in 1935 in relative obscurity. Ten years later, though, the Soviets found a German translation of Tsiolkovsky's book on spaceflight and rocketry at Heeresversuchsanstalt Peenemünde, the secret research center where the Nazis were developing the V-2 rocket under the guidance of Wernher von Braun. Now on display in a museum in Kaluga, the book has handwritten notes from von Braun himself on nearly every page.[1]

Einstein was a contemporary of Tsiolkovsky, although the two didn't know each other personally. Einstein would come to define gravity by 1915. From the Einsteinian perspective, to get into space we first need to climb out of the gravitational well created by Earth. Imagine, if you are at the bottom of a shallow well, you could easily toss a ball up and over the edge. The deeper the well, the harder you'd need to throw the ball to clear the edge. Earth's gravity creates a well, or indentation, in the fabric of space-time so deep that we need to throw a ball 1,800 meters per second, or Mach 5, to clear the atmosphere to a height of about 160 kilometers, or 100 miles. That's why rockets, which are essentially missiles, are used to get into space.

But *orbit* is more than just height. Indeed, going straight up is the easy part, about a fifth of the energy needed to get into orbit. Without constant velocity in a lateral direction, your ball would fall right back down to Earth. Remember the discussion about microgravity in Chapter 2. Any stationary object above Earth would fall back down; astronauts experience weightlessness because they are in a free-fall, with lateral velocity taking them forever tumbling over the horizon.

The Tsiolkovsky rocket equation, $\Delta v = v_{exh} \ln(M_0 / M_1)$, is what takes us into orbit and beyond. In this equation, the Δv, or delta-v, is the change in velocity from the launchpad to orbit, which is related to exhaust velocity (ν_{exh}), or how effectively a rocket uses propellant in a given gravity field, multiplied by the natural log function (ln) of the initial mass of the rocket with fuel (M_0) divided by the final mass of the satellite after the fuel is burned and booster rockets fall off (M_1). Now let's add some numbers. Satellites in low-earth orbit are moving at about 8 kilometers per second. This is incredibly fast, like flying from Los Angeles to New York in eight minutes, and lots of fuel is required to accelerate to that speed. This is the final speed needed to stay in low-earth orbit. Anything slower, and you'll fall back toward Spain; anything faster, and you'll fly off to a higher orbit.* To get a satellite to this speed, you need to endow it with a delta-v of about 10 kilometers per second, slightly more than the orbital velocity because of the drag caused by the very thin residual atmosphere. To reach a near-earth asteroid, you need a delta-v of 12 km / s; to reach the Moon, 14 km / s; to reach Mars, 16 km / s. So, you can see that just lifting off Earth to reach an orbit of about 100 miles or so requires about half the fuel as reaching Mars, thirty million miles away. (Traveling to Mars, you need fuel to slow down but not to maintain your speed of 16 km / s once you achieve that delta-v.)

Once you know the desired delta-v, the elements of the equation are rather fixed. The ν_{exh} is based on chemical efficiency of the rocket fuel, and for the most part, we are using the most powerful chemical fuel we have. The mass of the satellite placed in orbit is about 2 to 5 percent of the mass launched. The rest of the mass is rocket and fuel. Here's where the tyranny factors in.

*NASA maintained Transoceanic Abort Landing sites for the space shuttles, two of which were in Spain: Zaragoza Air Base and Morón Air Base.

Faster velocities require more fuel, and more fuel adds more mass, which requires more fuel, which adds more mass, which requires more fuel, which adds more mass, and so on. Switching to slower velocities once going fast requires fuel, too, because there's no aerobraking in space. You have to fire the engine in reverse. Going from any one kind of orbit to another—low-earth orbit, geosynchronous orbit, lunar orbit—requires an adjustment in the delta-v and an expenditure of fuel.

And while all that's just physics, there's also the economic equation. More fuel adds more money. More efficient fuel could lower the mass requirement, but experimental fuel costs more money to make or store compared to cheap rocket fuel, so you can't save much that way. Lighter materials could shave off some weight. But again, lighter materials that can withstand the rigors of launch are harder to craft and thus cost more money, so you can't save much that way, either. Will the tyranny ever end? As a matter of fact, if the Earth were just a little more massive, we'd never get off with our current blends of rocket fuel because the gravitational well would be deeper and no rocket would be large and light enough to hold the amount of fuel needed to lift it into orbit.

Many people are surprised to learn that the mass of a rocket on a launchpad awaiting blastoff is about 90 percent fuel and 8 percent metal casing to hold that fuel, with only 2 percent for the stuff we are putting into space: people or cargo. This current system—strapping ourselves to a couple of expendable roman candles with 500 tons of fuel and a sliver of cargo—is inherently expensive, costing more than $10,000 per pound of cargo. A gallon of water, $10,000. Breakfast, $10,000. A couple pairs of socks, $10,000. To create a space settlement, we'll need a lot of socks and other necessities. At the same time, access to space needs to be as affordable and dependable as access to the skies via airplanes or, as a historical analog, an immigrant's access to the seas via a ship. In the above scenario, your hundred-pound steamer trunk with all your belongings for the new world would cost $1,000,000 to bring along, and you haven't even purchased your ticket yet.

The cost won't drop significantly until, perhaps, we manufacture the necessities—air, water, food, clothes, shelters, and a big spacecraft—in orbit or on the Moon, thus eliminating the need to launch all this material from Earth. But do you see the catch-22? How do we create the infrastructure in space that will lower the costs of access to space if the cost is too high to launch the infrastructure? The rocket equation all but guarantees an astronomical expense to set up the infrastructure. You could bite the bullet and call it an investment in the future. But would this pay off? On one level, it makes perfect economic sense to establish Moon settlements to drive the space economy because the Moon is a big, low-gravity orbiting warehouse of fuel and raw materials ripe for the taking. Lunar resources could make Moon settlements profitable just as fish, fur, and timber made fortunes for European settlers in North America in the 1600s and 1700s. But even if you could magically go *poof* and make that infrastructure appear on the Moon, there are still no profits to be had because there's no market for our lunar dwellers to sell their goods, with access to space too expensive to be up there building stuff like orbital cities, massive solar arrays, and large spacecraft that would use these materials.

You can dream about living on the Moon or on Mars, with nifty domes and endless rows of hydroponically grown vegetables, but it is really just fantasy until we find a way to outsmart the rocket equation and to reduce the costs enough to make someone some kind of sure profit. Neil deGrasse Tyson explored this concept in his 2012 book *Space Chronicles: Facing the Ultimate Frontier.*

His stance then, and now, is that we don't have the right economic circumstances to motivate a settlement on the Moon, Mars, or anywhere else in the Solar System.*

Yes, It *Is* Rocket Science

How, then, can we lower the cost of access to space? There's little bargaining room with the Tsiolkovsky rocket equation. Tweak the fuel efficiency or tweak the mass. Rockets already are built with lightweight materials; and through rocket staging, empty canisters of fuel drop off to shed weight. Not much more mass to reduce. Not much more improvement can be made in regard to fuel, either. Many clever prototypes exist, such as ion drives, but these work only once you are in space. To fly out of the gravity well that is Earth, you need high thrust.

Nuclear could do it. Project Orion was a nuclear propulsion rocket system that could launch high mass with little fuel, essentially through the somewhat controlled explosion of nuclear fission bombs. Famed physicist Freeman Dyson co-led the project in the late 1950s. Nuclear fuel was a natural choice given its power, its novelty, its planned use to power submarines, and its general use by the military, which was calling the shots. The rocket would be propelled by the shockwaves that the bombs produced. On paper, the potential was stunning: a delta-v of hundreds of kilometers per second, powerful enough to get you to Mars in a week and to the nearest star, Alpha Centauri, in 150 years; and a design that allowed one-quarter of the mass to be payload, with the rest

*During his speech at the World Government Summit 2018 in Dubai on February 11, 2018, Tyson confessed that his original title for the book, rejected by the publisher, was *Failure to Launch: The Dreams and Delusions of Space Enthusiasts*. Ouch.

allocated to the rocket, the nuclear engines, and the necessary protection of the crew from the nuclear debris spewing out. The big problem with Orion was the high potential for harm from nuclear fallout should the rocket explode on launch. Dyson thought we could work through the problems, but the project was canceled in the early 1960s.[2]

With the nuclear option out, at least for launching, engineers long ago embraced chemical propellant. The thrust efficiency from such propellant requires a lot of fuel mass to lift a little rocket mass, a ratio of about nine parts propellant for one part rocket. The most common chemical propellants used today for launches are liquid fuels: a combination of liquid oxygen (LOX) and Rocket Propellant-1 (RP-1), a highly refined form of kerosene; a combination of LOX and liquid hydrogen; and nitrogen tetroxide (N_2O_4) and hydrazine (N_2H_4). In each case, we are combining something highly explosive with oxygen, which we need to carry along because there's too little oxygen in the upper atmosphere to sustain an explosion. These fuel blends have hardly changed much in fifty years. Of these fuels, the liquid hydrogen–oxygen mix has the highest exhaust velocity to plug into the Tsiolkovsky rocket equation, 4.4 km / sec.

(Note that, in the rocket equation, exhaust velocity is sometimes replaced by a term called specific impulse. The two are related: specific impulse, written as I_{sp}, is exhaust velocity (v_{exh}) divided by the force of gravity (g). Because g on Earth is about 10 m / sec^2, you'll see I_{sp} is roughly ten times less than v_{exh}. You can use the rocket equation for the Moon, Mars, or any celestial body simply by incorporating that body's gravitational force into the equation.)

There are experimental propellants with promise of high specific impulse but with hurdles to overcome. Trinitramide, $N(NO_2)_3$, discovered in 2010, might boost fuel specific impulse by 20 or 30 percent, but it is unstable and hard to handle. An aluminum-ice propellant called ALICE burns cleaner than other chemical propellants and is better for the environment, but the specific impulse doesn't outperform the alternatives. Metallic hydrogen is the most powerful rocket fuel in existence—now that it recently appears to have come into existence. Scientists at Harvard University led by physicist Isaac Silvera have created it, an advance that many are calling a game-changer.[3] If metallic hydrogen sounds exotic, that's because it is. It's hard enough to get hydrogen in a liquid form, let alone solid, let alone metallic. The stuff likely exists in the core of Jupiter under intense pressure, but no one has detected it directly. The Harvard researchers have created metallic hydrogen at very low temperatures by squeezing atomic hydrogen in a diamond anvil. There is speculation about how cheaply you could make metallic hydrogen or how safely you could store it; theory suggests it might be stable at higher temperatures once formed. All are in agreement, though, that metallic hydrogen propellant would knock the socks off of any other chemical rocket fuel. Metallic hydrogen is so powerful, in fact, that it would need to be cut with water to reduce the temperature when burned. According to Silvera, metallic hydrogen has a specific impulse of 1,700 seconds, or an exhaust velocity of more than 16 km/sec, four times higher than the best propellants now in use. This would dramatically reduce the cost of launch, depending on the ease of production and storage, the two big unknowns. The propellant could be powerful enough to launch a rocket with a single stage reducing rocket mass.[4] If you quadruple the specific impulse, you may be able to reduce fuel weight and increase the payload by a factor of a hundred.

And then there's fusion. Wouldn't that be nice? Fusion is the energy source of the Sun. High pressures and temperatures in the core of the Sun fuse hydrogen into helium, releasing copious amounts of energy in the process. Remember that *fission* is the splitting of an atom, a process in which atom nuclei break apart into smaller, lighter nuclei, such as in the reaction of uranium-235 breaking apart into krypton-92 and barium-141. (Add 92 and 141, and you get 233, not 235; that lost mass is converted into energy. Kaboom!)

Hydrogen fusion is far more powerful and less radioactive than fission. Moreover, the type of radiation is short-lived. The only problem is that, eighty years into the atomic age, we don't know how to generate fusion energy without the assistance of fission. That's what a thermonuclear weapon is, a fission bomb creating enough heat and pressure to generate a fusion reaction of even greater destructive force. Double kaboom!

Yes, fusion would provide a nifty rocket fuel, particularly for deep-space travel. You can throw away the Tsiolkovsky rocket equation at that point, though, because everything changes in a fusion economy. With unlimited, cheap energy, you could green the deserts, illuminate subterranean worlds, and create structures so high you could walk into space. This would be an event on par with the domestication of fire, the gods be damned.

We likely can't expect any revolution in rocket propellant in the next decade. And without improvements in propellant, not much can be done to the mass element of the Tsiolkovsky rocket equation. Rockets are just about as lean as they can be with a fuel base of hydrogen, the lightest of elements, and structures of lightweight yet durable metals. To lower the cost of access to space, entrepreneur Elon Musk and his company SpaceX are instead focusing on lean manufacturing of the rockets. The fuel, although we need a lot of it, is a small percentage of the launch cost, in the hundreds of thousands of dollars, not millions. The highest cost is the rocket itself. Those high-tech boosters that lift the rocket and pop off once their fuel is spent? They go for about $50 million a pop, so to speak. Musk has said a booster is about 70 percent of SpaceX's launch costs. Thus, SpaceX sees savings in the production and reuse of the rockets. The company has proven the technology, landing its boosters in a controlled fashion and reusing them within forty-eight hours.

Private Sector as the New Rocket Fuel

Rocket launches are akin to flying a 747 jet, destroying it after one use, and building a new 747 for the next flight. Ticket prices would tend to be high in such a scenario. And yet, that's the history of rocketry: rockets are missiles that carry cargo to space, as opposed to missiles that carry bombs to an enemy. NASA and the Russian space program have been married to the military since their creation, hence their dependency on missiles. President Dwight D. Eisenhower may have established NASA in 1958 to be distinctly civilian, but in actuality NASA was built on elements of the Naval Research Laboratory and the Army Ballistic Missile Agency. The latter employed Wernher von Braun, the captured German scientist who led Hitler's missile program before being transplanted to Alabama with his rocket designs on his knee.* Most NASA test pilots and astronauts came from the Air Force. During the 1960s, NASA was like the sixth element of the Pentagon; one look at those haircuts would prove this point. And the Pentagon doesn't launch missiles with the intent of retrieving them.

Reusing rocket parts, however, is not new. This was the concept behind NASA's space shuttles. The space shuttles were in-

*After World War II, von Braun and many of his staff were secretly moved to the United States as part of Operation Paperclip. They worked for the US Army on an intermediate-range ballistic missile program. NASA recruited von Braun as director of the newly formed Marshall Space Flight Center in Huntsville, Alabama, where he became the chief architect of the Saturn V rocket that propelled the Apollo astronauts to the Moon.

tended to glide back to Earth to be launched again. The twin boosters that lift the shuttles into space would fall into the ocean, where they would be recovered, refurbished, and reused. This looked good on paper in the early 1970s. But in reality, the boosters were so damaged with each launch that it became more expensive to refurbish them than to build new ones.* Worse, engineers were locked into a rigid shuttle design that forced them to reuse the boosters, adding considerable cost to each launch—in the half-billion-dollar range. More operational costs for the shuttle came as a result of Congress distributing pork and setting up contracts for shuttle parts in their own districts, from Florida clear across the nation to Washington, creating unnecessarily complex, expensive logistics. This was sold as a job creator, after all—that is, constituent jobs. All told, the final cost of the space shuttle program was about $1.5 billion per launch.[5] Even NASA now admits the program was a mistake.[6] NASA administrator Michael D. Griffin has stated that if the Saturn rocket program that took us to the Moon had been continued, it would have provided six manned launches per year at the same cost as the shuttle program. "If we had done all this, we would be on Mars today, not writing about it as a subject for 'the next 50 years,'" Griffin wrote in 2008. "We would have decades of experience operating long-duration space

*The main shuttle fuel tank, that big rust-colored one in the middle, was discarded just shy of orbit, where it would burn up upon reentry. Yet each tank had a greater volume than all of the ISS. Had NASA chosen to keep these hundred-plus tanks in orbit, collected and welded together in space, today we'd have a low-cost, ringed facility in low-earth orbit capable of generating artificial gravity and housing more than a thousand people. Scientist and author David Brin did the math and wrote about the concept in his fictional short story "Tank Farm Dynamo."

systems in Earth orbit, and similar decades of experience in exploring and learning to utilize the Moon."[7]

The space shuttle program was so convoluted that Soviet scientists assumed the shuttles were being designed to militarize space, because no sane government would invest so much money in such a flawed and impractical design simply in the name of science when there were far more efficient ways to get the desired scientific results. The space shuttle program petrified the Soviets.[8]

But I digress. SpaceX is aiming to avoid the sins of the NASA shuttle program by applying basic business savvy: create a logical supply chain and cut costs wherever possible. Lean manufacturing, vertical integration, flat management (i.e., open communication)—the hallmarks of Silicon Valley startups—this obvious approach has never been taken. United Launch Alliance, the consortium of Boeing and Lockheed Martin that produces both the Delta and the Atlas rockets, primarily for the US military, has had little incentive to cut prices for several reasons: no competition, a well-funded main customer, and shareholders who want to maximize profit. SpaceX and other startups want in on these lucrative contracts, an uphill battle considering the entrenched relationship among NASA, the military, Boeing, and Lockheed Martin. The newcomers need to not only launch more cheaply (for NASA and commercial satellites) but also deliver reliability, which the United Launch Alliance has done superbly.

SpaceX has the advantage of starting from scratch, to be nimble, whereas United Launch Alliance may be too bogged down in its old ways to innovate quickly, much like Ford and General Motors, who were taken by surprise by Toyota. SpaceX has the Falcon family of launchers and the Dragon family of spacecraft, as well as a heavy-load Mars-prototype rocket that, for several years, was known as the BFR. (BFR ostensibly stood for Big Falcon Rocket, although the F seemed to be used in other ways. Regardless, Musk changed the name in November 2018, calling the booster stage the "Super Heavy" and the upper, spacecraft stage the "Starship.") SpaceX cut costs on the Falcon with many clever approaches. Falcon 9's two stages, for example, both use the same kind of propellant, have the same diameter, and are assembled from the same aluminum-lithium alloy, which saves money on design and tooling for assembly and refurbishing. The Falcon is powered by Merlin rocket engines, which date back to the Apollo era, with space-proven reliability. Most rocket engines today use a showerhead-shaped injector plate to spray fuel and oxidizer into the combustion chamber; the Merlin uses something called a pintle injector, with a needle-shaped injector, which is both cheaper and less likely to cause combustion instability, the primary reason that rockets blow up on launch. Elsewhere, SpaceX has salvaged parts to reduce costs, literally using massive tanks and old railcars and the like sitting idle at NASA and military bases. The company builds or refurbishes most parts in-house to bypass the price gouging inherent in the aerospace market.

It could be that SpaceX has the new "right stuff." The company's derring-do in the business world is well documented. In one incident, the company was faced with needing an engine valve that the vendor said would cost hundreds of thousands of dollars and would require more than a year in development. SpaceX propulsion chief Tom Mueller found that to be outrageous and said so, and the supplier mocked him for his apparent naïveté. Nevertheless, Mueller's team made the part themselves and tested it for a fraction of the estimated cost. The supplier actually called back a few months after the initial discussion, hoping to strike a deal. Mueller explained, gleefully, to the supplier's absolute shock, that they were already done.[9]

Such a scenario has played out so often that SpaceX has learned to avoid space vendors. Another supplier wanted $3 million for an air-conditioning system to keep a satellite cool in the rocket's nose cone. Musk caught wind of this and asked how much such a system would cost in a house for a similar volume. The answer was a few thousand dollars, and that's exactly what SpaceX paid, with engineers doing some minor tweaking to fit it in the rocket fairing.[10] Then there's the heroic action on the battlefield. In 2010, on the eve of the Falcon 9's second launch, engineers found a crack in the nozzle, or skirt, of one of the engines. NASA's approach would be to postpone the launch for several months while engineers replaced the skirt entirely. At SpaceX, Musk called a meeting and asked what the impact would be from simply trimming the skirt just above the crack. He went around the table, person by person, to ask the ramification of this action for each one's particular part of the launch. The only drawback would be a little less performance from that one engine; but they had other engines to compensate. Within thirty minutes, the decision was made to trim the skirt. SpaceX flew a technician with a pair of shears from its California headquarters to Cape Canaveral, Florida, that evening; the technician cut the skirt, and Falcon 9 flew (successfully) the next day.[11]

The result of this approach? Cheaper launches. NASA paid SpaceX $1.6 billion for twelve cargo deliveries, considerably less money than the $1.5 billion it paid for a single space shuttle launch—and at least a third of the price of a United Launch Alliance (ULA) launch, although it's not quite an apples-to-apples comparison. The US Department of Defense (DOD) might follow suit. A 2014 US Government Accountability Office letter to the US Senate was highly critical of the Air Force's spending and ULA's launch prices, stating that "minimal insight into contractor cost or pricing data [has] meant DOD may have lacked sufficient knowledge to negotiate fair and reasonable launch prices" over the years with this monopoly.[12] Taxpayers may be becoming less tolerant toward wasteful government spending, even in space.

If SpaceX is spending $50 million per launch of the Falcon 9, as estimated, that's a hefty profit from the NASA contract. The profits lie in the number of launches and the reusability of the boosters, which Musk estimates could be flown a dozen times. SpaceX had eighteen launches of its Falcon 9 rocket in 2017 for paying customers, and in 2018 the company was averaging two launches per month, so money is coming in after years of testing and investments. One of those 2018 launches was the first successful test flight of its Falcon Heavy rocket; that's the one that placed a Tesla Roadster electric car with a dummy driver into orbit for reasons of pure showmanship. The plan was to send the Roadster in a path around the Sun that would sling the car to Mars, which is Elon Musk's destination for his future massive rocket (Super Heavy and Starship) to carry the first few travelers to Mars, with a longer-term goal of sending a hundred people at a time. The Falcon Heavy was a little too powerful, though, and it seems that the Roadster will overshoot Mars and head toward the asteroid belt.[13] (See the importance of the Tsiolkovsky rocket equation!) The Falcon Heavy comprises three Falcon 9 nine-engine cores. As of 2018, the Falcon Heavy was the most powerful operational rocket in the world by a factor of two, with the ability to lift into orbit nearly sixty-four metric tons. Only the Saturn V moon rocket, last flown in 1973, delivered more payload to orbit.

I don't mean to single out SpaceX as superior. There's lots of what I call "LUCA" ("loud, unfriendly counter-arguments") questioning the company's true spending, given that NASA so publicly failed to make reusability economical. And Musk may be a loose cannon bound to fail. In a live podcast on September 2018, he took a hit on what is said to have been a marijuana spliff that was offered to him; and while he later stated on the podcast that he doesn't care for marijuana and doesn't smoke it, the move ran-

Elon Musk's Tesla Roadster in space. This mind-blowing image is real, although the driver is a mannequin. SpaceX launched this electric car into space on February 6, 2018, aboard its Falcon Heavy rocket, an accomplishment that highlights private industry's growing interest in spaceflight. The car was intended to "drive" to Mars, but it appears to be on a trajectory past that, to the asteroid belt.

kled some NASA officials, who subsequently ordered an assessment of SpaceX's business culture.[14] But the point here is that several companies are vying to be leaders in rocket launches, and competition will surely be a path to cheaper access to space. Traditionally there has been no competition. For decades, only the United States and the Soviet Union launched rockets, both closely tied to the military. In 2010, there were only six primary rockets, properly called orbital launch systems, each either state-owned or nearly monopolized: the Ariane 5, manufactured under the authority of ESA and the Centre National d'Etudes Spatiales; the Proton-M, a Russian heavy-lift launch vehicle; the Soyuz-2, a smaller Russian rocket, which takes astronauts (including Americans) to the ISS; the Chinese Long March family; and the Atlas and Delta rockets operated by United Launch Alliance. Of all these, only the Russian Soyuz-2 and Chinese Long March 2F are used for humans. The proud United States has not had the ability to launch its own astronauts into orbit since 2011.

But the 2010s brought major change. Aside from SpaceX, three other companies are making serious headway into human spaceflight: Bigelow Aerospace, Blue Origin, and Virgin Galactic. What is particularly propitious is the synergy among these new companies. For example, Bigelow is in the space hotel business, and Blue Origin might fly you there, while SpaceX may land you on the Moon or Mars, and Virgin Galactic may take you around the world in two hours.

Then there are dozens of other companies filling in niches. New Zealand–based Rocket Lab is going small, with a tiny and relatively inexpensive rocket that places 225 kg loads into space, such as petite communication satellites that could provide the backbone for interplanetary travel. The company calls itself the "FedEx" of space delivery.[15] Indeed, there appears to be a healthy bifurcation evolving: large rockets are being designed to lift humans and heavy equipment, such as vehicles, diggers, and other necessities of exploration; and small rockets are proliferating to lift a new generation of nanosatellites, or nanosats, which are satellites less than 10 kg that, because of advances in miniaturization, can perform various tasks such as imaging and communications once requiring equipment a hundred times more massive. Even smaller satellites than these are being launched and deployed in swarms.⋆

⋆There are satellites called microsatellites, nanosatellites, picosatellites, and femtosatellites, each progressively smaller, but their names are more slang than a true metric categorization. The CubeSat is a type of picosatellite that's 10 × 10 × 10 cm.

A driving force behind smaller rockets is that they might find the market to launch weekly or even daily, lowering the launch price through economies of scale.

Collectively, these activities are part of the NewSpace movement, in which a diverse set of entrepreneurs and, increasingly, amateurs, share the overarching goal of reducing the price of access to space for pure commercial interests, independent of political motivations—in contrast to the "old space" partnerships of governments and military-oriented contractors. As summarized by Gary Martin, director of partnerships at NASA Ames Research Center, NewSpace represents "a turning point in the history of space exploration and development, the cusp of a revolution, [in which] new industries are being born that use space in many different ways. The established military industrial space sector is no longer the only game in town. Increased competition and new capabilities will change the market place forever."[16]

Price tags per launch can be confusing, though. Companies with smaller rockets often boast of launches costing only a few million dollars, compared with tens to hundreds of millions for larger rockets. Small rockets have small payloads, though. The convenient comparison is price per pound or kilogram placed in space, the cost-to-weight ratio. The company Rocket Lab's price of about $5 million per launch to carry 225 kg is about $22,000 / kg. The SpaceX Falcon Heavy can lift 64,000 kg for $90 million, which is only about $1,400 / kg. That's attractive, but there are fewer non-government customers needing sixty-four tons of cargo in space, compared with the hundreds of customers wanting to launch nanosats. For NASA and its rockets, the price has been historically about $10,000 per pound or $20,000 per kilogram. The space agency's goal, as articulated circa 1999, is to reduce the cost to hundreds of dollars per pound within twenty-five years and tens of dollars per pound within forty years.[17] How? It may be that NASA should rely on companies such as SpaceX and let economies of scale and savvy business practices lower prices even further. Zero tax dollars spent on development of basic launch vehicles would lead to significant savings that NASA could apply to other projects requiring a heavy launch—more profit for SpaceX or other vendors, more projects in space at the same net price. Win-win.

NASA clearly blew it with the development of the shuttle program and the ISS, so some folks have seriously questioned whether NASA, under the spending whims of the US Congress, can lead in aerospace technology development. The counterargument is that government, not business, must be the driver of high-risk, high-reward research—that is, the next generation of technology that private industry could then develop further in decades to come—as they are capitalizing now on the NASA advances from decades ago. And, in fact, NASA *is* working on next-generation launch vehicles. These include various X-planes to evaluate new aerodynamic concepts, such as the Lockheed Martin X-33, a suborbital spaceplane developed in the 1990s, and the Boeing X-37, also known as the Orbital Test Vehicle, a reusable unmanned spacecraft. Lots of lessons learned; not so many new things built as a result yet, though. Other NASA ideas for launch vehicles in development include, in order of feasibility given today's technology: air-breathing engines that rely more on atmospheric oxygen to burn fuel, boosting performance by 15 percent over conventional rockets; magnetic levitation to accelerate a vehicle along a track before launch; beamed propulsion from the ground to complement conventional fuel; electrodynamic magnetic tether acting as a thruster; pulse-detonation rocket engines, which use explosions as thrust through the nozzle; and exotic fuels such as fusion and antimatter.

Space Stations beyond the ISS

With cheaper access to space, however this is accomplished, far more people will visit, work, or live in low-earth orbit. Actually, we've been living in space continuously since the year 2000. But by "we," I mean only members of the 0.0000035714286 percent of the world's human population who have visited the International Space Station. Among these 250 or so people have been seven tourists. More are coming, as we will see. Not unlike the human presence on Antarctica, space habitation started as a source of national pride before evolving into a place to do science; as prices fall and safety is better assured, space will welcome millions of tourists.

A space station is an orbiting habitat launched via an unmanned rocket or built in space, with docking capability and room to accommodate a rotating crew that stays at the station for weeks or months at a time. NASA, the US Congress, and academics once seriously considered something much bolder than mere space stations—entire cities in space, to be built in the 1980s and habitable by the 2000s. Princeton University physicist Gerard O'Neill proposed immense, rotating structures two miles long with trees and flowing rivers and housing for thousands of people, whose primary occupation would be mining the Moon or otherwise maintaining gigantic solar arrays beaming unlimited energy to Earth. O'Neill spoke before the House Subcommittee on Space Science and Applications in 1975 and the Senate Subcommittee on Aerospace Technology and National Needs in 1976 about these O'Neill cylinders or O'Neill colonies, as they were later named. The concept of energy independence resonated well during the oil shocks of the 1970s, when energy prices soared. A major snag in O'Neill's plan, however, was that it relied on the assumed cheap access to space via the space shuttle, which hadn't yet flown.

An artist's impression of an orbiting torus, or city in space. Such massive, rotating structures could provide living space for hundreds of thousands of people with comfortable temperature, artificial gravity, ample room for growing food, earthlike day-night cycles, and unlimited solar energy, all in the vicinity of the Earth and Moon.

By the 1980s, when oil prices fell and shuttle prices went through the roof, the concept of building low-orbit suburbs for solar energy workers was quickly abandoned. These large, orbiting spheres remain an exciting option for space settlements, however, because they can provide artificial gravity and other earthly niceties.

The Soviets led the development of space habitats in the 1970s. They ran two programs simultaneously, one for science and the other for military purposes. To the outside world, these were known as the Salyut stations, and there were seven in all, five of which were successful. Only decades later did the world learn that three of the stations—Salyut 2, 3, and 5—were actually internally designated as Almaz 1, 2, and 3, part of a secret military program to test space reconnaissance tactics. The Soviet success in launching so many stations was a result of a "space race," but not with the United States. This was a heated, internal competition between Kerim Kerimov, who designed the Salyut stations, and Vladimir Chelomey, who pushed for military stations.[18]

Salyut 1 launched in April 1971 and remained in orbit for 175 days, hosting a three-man crew for 23 days. The US answer to Salyut was Skylab, launched in May 1973 on a Saturn V, the last time that iconic rocket was used. Skylab was about three times larger than the Salyut stations, with a main habitat area that was 48 feet long and 21.6 feet wide. The station was a relative success, with three visits each with three astronauts, lasting 28, 59, and 84 days, respectively. But Skylab was plagued by problems. Its micrometeoroid shield tore off during launch, taking a main solar array with it. The first crew needed to fix the damage before they could move in, a major and unprecedented feat that nevertheless frustrated the astronauts. As NASA documented, "The astronauts were venting their frustration with four-letter words, while Houston repeatedly tried to remind them that communication had resumed," and their cussing was being broadcast to the world.[19] And while the amount of scientific data collected was impressive—novel Earth, solar, and comet observations and the first long-term studies on weightlessness—the constant repairs grew tiresome. Skylab's third and final crew complained bitterly of the tasks assigned to them and, by some accounts, contemplated a mutiny. Said crew member Edward Gibson to mission control, "I personally have found the time since we've been up here to be nothing but a 33-day fire drill. . . . I've been engulfed in building blocks rather than being concerned with the quality of the data."[20]

Skylab was abandoned in 1974. The space shuttle was to be used to boost Skylab's orbit, but development of that was so far behind schedule that Skylab's orbit slowly decayed to a point of inevitable reentry. The whole thing came tumbling down to worldwide mass hysteria in 1979. NASA had limited control over the descent. Despite assurances to the public that the station would mostly burn up over the ocean during reentry, NASA engineers miscalculated, and large chunks sprinkled Perth, Australia. Remarkably, and quite luckily for NASA, no one was hurt. In fact, one young Australian named Stan Thornton found a chunk and immediately flew off to the United States to claim a $10,000 prize offered by the *San Francisco Examiner* for a piece of Skylab.

The Soviet answer to Skylab was Mir, launched in 1986 and assembled and expanded in modular fashion over the next ten years. Cosmonauts who visited set all kinds of records for duration in orbit and length and complexity of extravehicular activity, or EVAs. Several inhabitants spent more than a year on Mir. Mir also became the blueprint for the ISS in terms of its modular design, life-support systems, hygiene facilities, food and drink provisions, sleeping quarters, and science experiments.

Built between 1998 and 2011, the ISS is a partnership among five participating space agencies: NASA, the European Space Agency, the Japan Aerospace Exploration Agency, the Canadian Space Agency, and Roscosmos (Russian Federation). The cost to construct the ISS was $100 billion, of which NASA paid $75 billion; and the cost to run the ISS through its proposed lifespan to 2024 is $3 billion to $4 billion a year, an estimate the NASA Office of Inspector General calls "overly optimistic."[21] The ISS is actually irrelevant in our discussions about living or working in space, because in no plausible scenario will more than a few people be living or working in an ISS-like environment—that is, in zero gravity. All space settlements will occur with artificial gravity or, if it's enough, the natural gravity of the lunar or planetary body of interest.

The ISS admittedly has an element of awesomeness to it. But it was not built as a prelude to space colonization. Rather, the ISS was envisioned as a microgravity laboratory, primarily for learning about physics and chemistry in microgravity and not so much about human health. Compared with the best earthbound labs—US national labs such as Los Alamos and Lawrence Berkeley, Europe's CERN particle accelerator, UK's Laboratory of Molecular Biology, or the legendary Bell Labs—science results from the ISS are modest indeed. Most of the science on human health conducted on the ISS is a confirmation of findings from Mir experiments—which can be summed up as microgravity being bad for human health, so bad, in fact, that the only cure is to get out of microgravity as quickly as possible.

Technological gains have been made on the ISS, to be sure. Many of these, however, such as docking and maneuvering, are a refinement of what was learned with the earlier space stations. NASA has improved methods for water and air filtration on Earth, but these are spin-off technologies resulting from a necessity of life support on the ISS, not ISS microgravity research in and of itself. At its best, the ISS has been a place to train astronauts for work in zero gravity, to test new technology in zero gravity, and to practice receiving cargo from private industry, all in hopes of learning how to better build things in space. True commercial research and development in zero gravity has always eluded the ISS because of its expense.

China is planning its own large modular space station for low-earth orbit, to be assembled by 2022, the third tier of its Tiangong program. China launched prototypes called Tiangong-1 and

Tiangong-2 in 2001 and 2016, respectively. Although rarely reported in the Western news media, the program has been a success. Tiangong-1 received astronauts, or taikonauts, twice before being deliberately deorbited in April 2018; Tiangong-2 also has received taikonauts and may dock with the planned modular space station, the core module of which will be called Tianhe. The full space station will be about the size of Mir, or a fifth of the ISS. As such, China is reinventing the wheel, entering the league of nations with space stations developed primarily for national pride.

The first steps toward space settlements beyond low-earth orbit, however small, may be NASA's planned Lunar Orbital Platform-Gateway (LOP-G), formerly known as the Deep Space Gateway. With the LOP-G, NASA is eyeing an orbit in cislunar space. Why not go straight to the Moon and bypass the space station thing? And Mars? What happened to Mars? Aren't we supposed to be going to Mars? As contrived as it may sound, the LOP-G is intended to be a gateway to both the Moon and Mars. From the LOP-G, a station begging for a better name, a small crew could operate a robotic mission on the lunar surface and even pilot landers on the Moon for human exploration in the late 2020s.[22] NASA acting administrator Robert M. Lightfoot Jr. described the station as like a "cabin in the woods" that international partners could stop by to use in their exploration of the Moon.[23] He envisions NASA using it to scout the Moon for ideal permanent base locations and prospect for minerals.

The LOP-G is not the ISS 2.0. For starters, it's much smaller, similar to Mir with just a single habitation unit, airlock, and a power and propulsion unit. It's cheaper (we hope): about $2 billion compared with the ISS $100 billion construction cost. And it's less about PR and camera-posing astronauts and more about single-focused tasks requiring no more than a month's stay. But the LOP-G is not without its many critics, including former NASA administrator Michael Griffin, who has called it "stupid."[24] At issue is the fact that we could operate lunar-based robots from Earth just as easily as from the LOP-G; the signal delay is only about a second. And docking to the LOP-G before descending to the Moon offers no advantage; your primary craft could just as easily maintain its own orbit, as was shown in the Apollo missions. At best, the LOP-G could be useful as a fuel depot with fuel extracted from the Moon. This would result in massive cost savings for trips to the Moon, the asteroids, and Mars because spaceships wouldn't need to carry all the needed fuel from Earth. And yet, NASA is building the LOP-G in advance of any lunar mining infrastructure and incorporating it in their plans for a human return to the Moon, which have yet to be articulated.

Nevertheless, at a symposium held in March 2018 in Washington, DC, called "Return to the Moon: A Partnership of Government, Academia, and Industry," many participants—including Harrison "Jack" Schmitt, the Apollo 17 astronaut who was the second-to-last person to leave the Moon—were galvanized by NASA's plan with the LOP-G, more so than they have been for the dozens of other "Return to the Moon" symposia they have attended over the decades. This time is different, they told me. The LOP-G places us at the precipice. Moreover, private companies such as SpaceX and Blue Origin can carry the cargo for NASA to the LOP-G, reducing the price by an order of magnitude. ESA and Japan are working complementarily with NASA on lunar mission designs that might utilize LOP-G. And, perhaps most important, China is the new Soviet Union providing additional incentive. Alluding to a new space race, Schmitt said that China's lunar ambition is providing a "geopolitical imperative" for the United States to return to the Moon via the LOP-G.

Space Hotels: A Weekend Getaway Like No Other

On April 28, 2001, US businessman Dennis Tito arrived at the ISS aboard a Russian Soyuz TM spacecraft to spend nearly eight days in orbit. He paid the cash-strapped Russian Federal Space Agency $20 million for the thrill, effectively becoming the world's first space tourist. NASA was furious about this plan and refused to train Tito at the Johnson Space Center, but they couldn't stop the visit. Six other multimillionaires made similar trips to the ISS between 2002 and 2009, until Russia halted the program for logistical reasons because, with the retirement of the NASA shuttle program, the Russian Soyuz vehicles are the only way to get anyone to the ISS, and space is limited. (NASA would soon be paying Russia $80 million for a ticket.)

Nevertheless, the era of space tourism is upon us. Russia has stated interest in resuming tourism flights. And even some NASA astronauts, appalled by the idea at first, are warming up to the reality. Astronaut Michael Lopez-Alegria, for example, expressed reluctance over hosting space tourist Anousheh Ansari, the Iranian American cofounder of the technology company Prodea Systems. After Ansari's 2006 visit to the ISS, however, Lopez-Alegria had a change of heart. "If that's the correct solution [accepting tourists for large sums of cash] . . . then not only is it good from the standpoint of supporting the Russian space program, but it's good for us as well," he said.[25]

And NASA itself is on board with the idea. In June 2019, NASA announced its plan to commercialize the ISS—again.[26] NASA has never had much commercial interest in the ISS, with companies not seeing the worth of the pricy investment. In this latest announcement, though, NASA stated that it would welcome space tourists—or, as NASA carefully worded it, "private astronauts," which is a euphemism for multimillionaire thrill-seekers. The price would be $55 million for the rocket trip and about $34,000 per day for life support and luxuries such as water and air. The offer is contingent on NASA developing or otherwise securing a rocket to take humans to the ISS, of course. Bigelow Aerospace and SpaceX have voiced their interest in getting "private astronauts" to the ISS for a tidy profit. So, the NASA plan strikes at the heart of the matter in indirectly acknowledging that the first real money to be made in the realm of human spaceflight will be tourism. This may be the only way to keep the ISS in orbit once Congress tells NASA to pull its funding for the station in 2025, as is planned.[27]

Bigelow Aerospace, through its subsidiary Bigelow Space Operations, formed in 2018, is pioneering a new market of lightweight, durable, inflatable modular habitats for space hotels or for expanding existing space stations. Far from hot air, the company is built on solid funding and expertise. The founder is billionaire Robert Bigelow, who made his fortune with the Budget Suites of America hotel chain. He's a hotel man, and his long-term goal is to bring some version of Budget Suites to space. In 2016, Bigelow Aerospace successfully placed its Bigelow Expandable Activity Module (BEAM) on the ISS, to be tested for safety for a year. That went well; micrometeoroid damage was minimal, and radiation levels were comparable to those on the rest of the ISS. In 2017, NASA announced plans to keep BEAM inflated through 2020 as a storage unit for various ISS junk.[28]

Bigelow Aerospace delivered BEAM to the ISS aboard a SpaceX Dragon cargo spacecraft, a demonstration of the new interconnectivity among space ventures. Also of worthy note is the fact that Bigelow Aerospace purchased the rights to the patents of the inflatable habitat technology from NASA after the US Congress canceled the program in 2000. NASA had been working on

The Bigelow Expandable Activity Module (BEAM). Lightweight, inflatable structures such as these may serve as the first space hotels. BEAM is the brainchild of Robert Thomas Bigelow, founder of the Budget Suites of America and, more recently, Bigelow Aerospace. Among the first visitors to a BEAM could be movie and music video crews. One BEAM was successfully attached to the International Space Station in 2016.

inflatables since the 1960s and, in the 1990s, was developing a massive inflatable called the TransHub for roomier accommodations on the ISS. Lightweight, twice the diameter of the ISS habitats, and made of durable Kevlar and Nextel fibers to withstand micrometeoroid impact, TransHub was designed with a mission to Mars in mind. Congress's canceling of the program came as a shock to the TransHub team.[29] But unintentionally, the move proved to be successful. Bigelow Aerospace, immune to Washington bureaucracy and the whims of Congress, brought the technology to fruition within six years. TransHub designer William Schneider compares Robert Bigelow to Howard Hughes—motivated, daring, and involved in all aspects of the engineering.[30] True, Robert Bigelow believes space aliens live among us, but such beliefs may drive his passion to find ways for millions of people to leave the confines of Earth.[31] The market is there; the technology is almost there. What will it look like?

First, let's define a few terms of the places we will visit. Low-earth orbit (LEO), where most of the tourism will initially take place, is between about 100 and 1,200 miles (160 to 2,000 kilometers) above Earth. The ISS, the Hubble Space Telescope, and remote-sensing and spy satellites are all in low-earth orbit. Medium-earth orbit (MEO), home to a multitude of GPS and communication satellites, is about 1,200 to 22,000 miles (2,000 to 36,000 km) above Earth. A geostationary orbit, where a satellite is always in the same spot above Earth along the equator as the Earth makes its twenty-four-hour rotation, is at an altitude of 22,000 miles, or 36,000 km. This is also the altitude of a geosynchronous orbit, a twenty-four-hour orbit that can be in any direction around globe, not only above the equator. A multitude of communication satellites are in geosynchronous orbit; lots of military and weather satellites are also there.

There are numbered sweet spots, called Lagrangian points, at which the gravity between two objects, such as the Earth and the Moon, or the Earth and the Sun, is balanced enough to offer a nearly free ride for any object placed there. Numerous astronomical observatories are in such locations. Physicist Gerard O'Neill envisioned building orbital cities at Earth–Moon Lagrangian points 4 and 5, called L4 and L5, which are along the path of the Moon's orbit, either 60 degrees ahead or behind it. An object placed there would remain there indefinitely, without needing to expend fuel, much as the Moon goes around the Earth. That's 240,000 miles (384,400 km) away, essentially the distance from Earth to the Moon.

Low-earth orbit will be the spot for orbital hotels not only because it is easier to get to compared with other orbits but also because it is within the safe confines of the magnetosphere. Recall, the magnetosphere blocks a significant portion of solar and cosmic radiation. The magnetosphere isn't a giant, fixed, perfect sphere around Earth; it dips precipitously near the Earth's poles. But, roughly speaking, the closer you are to Earth, the more protected you are, and anything beyond geosynchronous orbit is beyond the sphere of protection. And space is space. If you're not on the Moon or Mars, what's the difference, really, between a hotel in LEO or MEO? The view is essentially the same—that view being awesome.

Space hotels will certainly lack the comforts of earthbound accommodations, at least at first. There will be no mint on the pillow, essentially because it could float off in microgravity. Amenities won't matter much, because all that the tourists will desire is the experience of living in microgravity for a few days, long enough to enjoy the sensation before the frustration and adverse health effects set in. The ISS itself could be a space hotel. If NASA leaves in 2025, as planned, it could be that Bigelow Aerospace or some other player—and there are many, in various stages of development, some with business plans and investors, others with only PR teams—would fill that vacancy and convert the US modules to hotel rooms. Renting a room on the ISS, as opposed to using it for some pharmaceutical or other science experiment, may be the only profitable activity once government pulls out. Of course, four other governmental bodies aside from NASA are involved with the ISS, none too happy about the space station's fate being dictated by the United States.

Private industry also has plans for free-standing (or free-flying) small hotels the size of a school bus for wealthy clientele to visit for about $10 million, roughly $1 million per night, whatever "night" might mean. Bigelow Aerospace hopes to launch, by the early 2020s, two inflatables, called B330s, each fifty-five feet long and twenty-two feet wide, which when linked together will created the first space hotel, with double the cubic capacity of the ISS.[32] Aside from the thrill-seeking multimillionaires, there surely could be movie and music video directors with their actors and pop stars eager to film in microgravity and willing to pay the few tens of millions of dollars to visit LEO. The opportunity could prove to be an investment that pays lucrative dividends, considering the popularity of shooting footage in orbit with no need for special effects.

The lower the price per unit mass of cargo becomes, the more complex hotel construction can be. Private companies are planning larger rotating structures, similar in design to Space Station V featured in the movie *2001: A Space Odyssey,* a rotating torus connected via spokes to a central hub. Gravity in the torus would seem normal as a result of centrifugal force. Traversing the spokes to the central hub, you would experience a diminishing gravitation force until you reach the zero-gravity environment of the hub. The hub would be the microgravity "play area," while the torus would be the place for sleeping, dining, gambling, or other kinds of typical resort entertainment.

These projects would take decades to complete. One convenience of space construction, though, is that massive structures can be assembled in modular fashion and become habitable even while they are growing, unlike a traditional high-rise hotel or office tower. Indeed, if you look closely at the fictional Space Station V, you will see that the second torus is under construction. The concept isn't at all fictional. The main limiting factor for building massive space structures is the prohibitive cost of lifting mass off of Earth. However, robotics and 3-D printing are advancing to a point that, once launch prices drop, construction could commence. Remote-control robots already perform most of the mining in Sweden; construction in space similarly could be directed by operators with joysticks on Earth.

Tourism and entertainment activities in low-earth orbit, driven by the likes of Robert Bigelow, are poised to be the first profitable *human* activities beyond Earth. Nothing else done with humans in space has ever been profitable. Communication satellites are extremely profitable but don't involve humans sitting inside them. The Moon landings cost billions of dollars and were done for national pride and military concerns. The ISS costs billions of dollars and involves just a smidgeon of science; any positive return of investment is highly questionable. But space tourism is uniquely positioned to drive the commercialization of human spaceflight in a manner that will make it cheaper and safer.

This commercialization can run parallel with manufacturing interests, as well. Cheaper space access will allow for the creation of orbiting factories for making unique commercial and industrial products in a gravity-free, vacuum environment. This could include the aforementioned protein crystals attempted on the ISS and also films and polymers, whose profitable production is stifled solely by the expense of placing materials in orbit.

NASA and its Soviet counterparts have perfected the art of moving through outer space. And while we can be glib about their poor decisions and cost overruns, the truth is that private industry would be in no position to build rockets and make orbiting hotels without the pioneering work of the US and USSR governments.

Sex in Space

OK, I'll talk about it. Sexual intercourse in space is a fascination among many. You can imagine that any couple visiting an orbiting hotel will want to give it a whirl. In reality, sexual intercourse in zero G might be more complicated than you think because of that sexy little topic known as Newton's third law: action-reaction. Any thrusting action by one body causes the other body to go flying in the opposite direction. So, at a minimum, you would need to be strapped together with your partner and secured to a wall so that both of you don't go flying together. There exists an amateur prototype spacesuit called the 2suit that would bring you in close contact with your partner. But a suit, or any clothing, is likely a bad idea because you will get hot and, in microgravity, your sweat doesn't evaporate nicely. It pools where it is excreted, and quickly becomes a slimy mess. Maybe none of this is too bad, so far. You might actually get better with practice.

Next challenge? Blood flow. In microgravity, the heart isn't pumping as much blood towards the genital area in times of, well, need. For men this results in a smaller, weaker erection; for women, less arousal and internal lubrication. Compounding this for men is the general lower level of testosterone experienced in microgravity. Also, anyone newly checked in to a space hotel likely will be tired, short of breath, and nauseated. It would be the excitement and novelty of the sexual experience that propels the couple forward in their mission. A female–female experience might be more satisfying, compared with male–female or male–male, as concluded by Paul Brians in his short story "The Day They Tested the Rec Room."[33]

No astronaut or visitor to the ISS has yet to have had sex up there. How can I be so sure? Because they or their crewmates would have talked about it. There's little privacy for one person on the ISS, let alone two. Some people talk of the challenges of sexual intercourse in space as if the future of humankind was in jeopardy. An entire book, *Sex in Space,* concerns itself with the topic. But this is utterly a nonissue. Should couples want to try sex in an orbiting hotel, there's no harm in that. Any deeper voyage in the universe, however, would be done in an environment of artificial gravity.

Real Shuttles to Orbit: Spaceplanes

So, how do you get to a space hotel—or space factory? You fly up in a spaceplane, of course. In the large space resort described above, spaceplanes would land in the central hub, much like a military jet lands on an aircraft carrier.

I'm getting a little ahead of myself, though. For now, the only way into space is via a rocket launch. That's how the first space tourists reached the ISS, and that's how the first visitors to space hotels will reach their destinations. Companies already are booking rocket flights and rooms with expectations of entertaining visitors by the early 2020s. By mid-century, infrastructure might be in place to enable day trips into cusp of space, for work or for fun. But rather soon, spaceplanes will be available, albeit for a high cost.

Flying straight into space from the ground in a vehicle resembling a jet plane, as we see in movies such as *Star Wars,* is difficult without nuclear propulsion. We would need at least three kinds of engines: one primed for the lower, oxygen-rich atmosphere; one efficient in a high-altitude, low-oxygen environment; and one for the vacuum of space. Such engines exist, but not all in one vehicle. Jet engines are remarkably efficient at altitudes above about

40,000 feet (7.5 miles, 12 kilometers). The engines take in oxygen-rich air, funnel it to the combustion and turbine chambers, and produce exhaust that propels the plane to a seemingly fast but relatively low velocity of under Mach 1. At higher altitudes, where the air is thinner, we need engines such as ramjets or scramjets, which rely on higher speeds to funnel oxygen into a combustion chamber. These don't work well at lower speeds—that is, the speed of the plane during takeoff. But once they get going, they can reach hypersonic speeds of Mach 6, albeit far short of the Mach 22 needed to get into orbit. The fastest plane to date was the unmanned NASA X-43, which reached Mach 9.6 with scramjet technology—delivered to high altitude by a conventional jet plane. To reach orbital velocity with your spaceplane, you still need a rocket engine on board.

Europe has a forward-thinking but poorly funded spaceplane concept called Skylon, a single-stage-to-orbit vehicle contemplated by the British company Reaction Engines Limited, in development since the 1980s. The plane would utilize a hybrid scramjet-rocket engine called SABRE, or synthetic air-breathing rocket engine. This engine would work by drawing oxygen from the atmosphere for the oxidizer up until about Mach 5 and then switching to stored rocket fuel (hydrogen plus oxygen) for the extra thrust needed to go faster and higher. Engineers, however, need to overcome the issue of drawing in oxygen at high speeds, above Mach 2, while keeping the intake cool, the long-standing stumbling block that has stymied the creation of spaceplanes all these years. The company needs billions of dollars to fund this project to its completion, but funding has fallen far short of that, with money only trickling in for concept testing from ESA and, more recently, US Defense Advanced Research Projects Agency (DARPA).[34] This lack of funding gets back to the concept of necessity. In the 1960s, the space race provided the necessity for funding rocket development, and what a cost that was! With Apollo-like funding from the start, Skylon would likely be a reality today. But there's no necessity for funding it.

Spaceplanes like Skylon, and whatever NASA is dreaming up, would complement conventional rockets and stick to very low-earth orbit. They would be ideal for ferrying people but not for heavy cargo. At least five types of spaceplanes have flown successfully. None really meet the bill as a true plane that takes off horizontally and flies into orbit, as you might envisage. For example, some characterized NASA's space shuttles as spaceplanes, but they were really just gliders that needed rockets to carry them into space. The Soviets had a nearly identical concept, called the Buran, which flew only once, in 1988. It survived that flight easily enough but was crushed to pieces in a bizarre hangar accident years later. The Boeing X-37, operated by the US Department of Defense with NASA participation, is like a scaled-down space shuttle, first launched atop an Atlas 5 rocket in 2012. The X-37 flew five times between 2012 and 2017, including once with a boost from a Falcon 9. This is a classified and presumed military project; there's no public information about the purpose of these unmanned flights, some of which have lasted well over a year.

The first real plane to almost reach outer space was the US Air Force's X-15, a hypersonic rocket-powered aircraft, flown in the 1960s, that took off horizontally and exceeded an altitude of fifty miles, qualifying the pilots as astronauts. SpaceShipOne was the first private spaceplane to reach space, in 2004, crossing the Kármán line, the internationally recognized boundary at an altitude of 100 kilometers or 62 miles. That's shy of orbit, but SpaceShipOne did win the $10 million Ansari X Prize as the first nongovernment organization to launch a reusable manned spacecraft into space

Virgin Galactic's SpaceShipTwo spaceplane (central fuselage). This spaceplane is lifted to an altitude of 15 kilometers by the White Knight Two cargo plane and let go; it then climbs to 110 kilometers, 10 kilometers higher than the Kármán line, which marks the boundary of space. Virgin Galactic plans to operate a fleet of five SpaceShipTwo spaceplanes that would take six passengers at a time on a suborbital flight for $250,000 per ticket.

twice within two weeks. SpaceShipTwo has shown great promise. This would be a spaceplane launched midair after being detached from a jet-powered cargo aircraft. Virgin Galactic articulated plans to operate a fleet of five SpaceShipTwo spaceplanes and has been accepting bookings for flights costing passengers $250,000 a ticket.[35] Tragically, the first of these planes crashed during a test in 2014, killing one pilot and seriously injuring the other. Richard Branson, chairman of Virgin Galactic, promised flights as early as 2010. The crash has pushed back the first commercial flight to 2020 at the earliest, an estimate based on their successful test flights of December 2018 and February 2019.

Jeff Bezos, founder of Blue Origin, has promised much the same for suborbital flights as a gateway to orbital flights, once he is sure his company's liquid-hydrogen-burning BE-3 engine is reliable.[36] He's shooting for the early 2020s. When the planes do take off—and you can view this as a commercial space race—six passengers at a time will be treated to several minutes of microgravity while viewing the curve of the Earth set against the blackness of space 100 kilometers in the air.

Despite the schedule slips, this is real. The activity of Virgin Galactic prompted the Federal Aviation Administration (FAA) to issue regulations in 2006 establishing requirements for crew and passengers involved in private human space flight.[37] You know the drill: keep your seatbelt fastened; no smoking in the lavatories. The US Congress mandated regulations in the Commercial Space Launch Amendments Act of 2004. Recognizing that this is a fledgling industry, the law allows for a phased approach in regulating commercial human space flight, with regulatory standards evolving as the industry matures. Most important, the act didn't require safety levels expected from air travel to apply to space travel; passengers on suborbital or orbital flights would need to sign an informed consent document stating they understand the risks.

Some may find it surprising that you can reach a LEO hotel on a spaceplane rather quickly, in less than an hour: it is only about a hundred miles away from any point on Earth. The hard part will be getting to the airport—or, more accurately, the spaceport. Virgin Galactic intends to fly out of the Mojave Air and Space Port, located in Mojave, California, or Spaceport America in New Mexico, both of which are deliberately not near populous areas. That reason is twofold: rockets and spaceplanes are still a bit dangerous and could explode; and they are loud. Most folks dislike living near airports because of the noise, but a spaceport would be much worse. A jet launch is about 150 decibels; a rocket launch is about 200 decibels. (A decibel is the unit on the logarithmic scale of sound, so an increase of 50 decibels is 10^5 or 10,000 times louder.) A spaceplane presumably would be as loud as a jet during takeoff but, reentering the atmosphere at several times the speed of sound, it will produce a sonic boom as loud as a thunderclap. The reentry of the X-37 could be heard all across east central Florida early in the morning, waking residents. Once a year or once a month may be bearable, but hundreds of times daily, akin to jet traffic, would be maddening.

Aerial view of Spaceport America. Situated in the Jornada del Muerto desert basin in New Mexico, the complex is the world's first purpose-built commercial spaceport, completed in 2011. The site can accommodate vertical- and horizontal-launch aerospace vehicles. Spaceports must be built in remote locations because of safety and noise concerns. Current tenants of the spaceport include Virgin Galactic.

Noise from spaceplanes and space-bound launches in general poses a significant problem to overcome but a necessary one to resolve if we are to truly support a space-based population and economy. Let's be clear on the numbers: to bring thousands of visitors to a space resort each year, you would need weekly space flights. To provide access to space to millions of people yearly—still only 1 percent of the human population—you'd need at least fifty launches and landings each day. How do we handle this noise?

The Mojave Air and Space Port holds the distinction of being the first US facility to be certified as a spaceport by the FAA for horizontal launches of reusable spacecraft, four days before the SpaceShipOne flight in 2004. Between then and 2018, nine more spaceports have since been certified by the FAA for commercial use, from the cleverly named Mid-Atlantic Regional Spaceport (MARS) on Wallops Island, Virginia, to clear across the United States to the Pacific Spaceport Complex–Alaska, on Kodiak Island. As of 2019, no other nongovernmental spaceports exist globally, but plans to build them have launched in Scotland, Sweden, densely populated Singapore, and elsewhere. So rapid is the development of spaceports globally that any list I could provide in this book would be outdated within a year.

The first generation of spaceplanes will be all about the flight, a circular trip to space and back for the pure sensation of weightlessness and seeing the Earth from afar. These planes can be modified, however, to take you to any destination on Earth in under three hours. Virgin Galactic hopes to fly from London to Sydney in two and a half hours—a flight that normally takes twenty-two hours. New York to Hong Kong? Only two hours. The effect this would have on world travel would be stunning. One could hop across the globe for a meeting and be back for dinner, admittedly at a cost. There is precedent for this. The Concorde was a supersonic jet that could fly from Paris or London to New York in just over three hours, more than twice as fast as any other plane. Known for speed, not luxury, the Concorde fetched a ticket price of $8,000 in the 1990s, which would be about $13,000 in 2019. Reducing twelve- to twenty-hour flights to just two hours could similarly fetch a handsome ticket price, but a fare easily within reach of the wealthy. The one snag could be the extra time it takes to get to a remote launch site, in a desert or out at sea.

Beyond Rockets: Skyhooks

Spaceplanes might not get you all the way to the Moon, but they can get you high enough to get hooked—onto skyhooks. A skyhook is a tether system in space that can whip cargo into orbit at a fraction of the cost of a rocket launch. Imagine a large platform orbiting above the Earth, as a satellite would, only with a cable and hook dangling from it. If a spaceplane could deliver a payload to that hook at just the right time, when the hook is passing at orbital speed, the payload would be swept up into a velocity close to that of the hook. Now that payload can get flung into orbit without the need for rockets. That's the simplest concept of a skyhook, and it's technologically feasible today. This basic version—a dangling line, hook, or magnet—is called a nonrotating skyhook. The concept actually was tested successfully in the 1950s with traditional aircraft at an altitude of about 1,000 feet, which might be easier to visualize. It was called the Fulton surface-to-air recovery system (STARS), developed by the US Central Intelligence Agency (CIA), air force, and navy for retrieving persons on the ground. In this scenario, the person to be recovered receives a supply drop comprising an uninflated balloon, a canister of helium, a protective suit, and a harness. The person inflates the balloon and rises to a suitable altitude, where a plane swoops in and hooks the apparatus. The person then, perhaps surprisingly, gently and gradually reaches the speed of the airplane and, once the line is taut, is reeled in like a fish.[38] This is not known to have ever been used in a real-life emergency, but such are the ways of the CIA, air force, and navy—neat ideas funded and tested, just in case.

With a nonrotating skyhook, the hook is a long line dangling at the edge of the atmosphere, suspended from a platform in LEO moving at orbital speed. Stratospheric balloons or a spaceplane could lift cargo or even a capsule with human passengers to the passing hook, which then could be pulled up to the space-based platform. A more dynamic system is a rotating skyhook. Envision a drum major's twirling baton. Now scale that up to a thick four-mile-long cable twirling like that in space, moving at orbital speed, with the twirling perpendicular to the orbital direction so that the cable tip dips into the upper reaches of the atmosphere and then back into space. Cargo caught by the cable would be swept up, attain orbital velocity, and be let go when the cable tip reaches the point 180 degrees opposite of where it hooked the cargo.

In the 1990s, the aerospace company Boeing tested a skyhook protocol called the Hypersonic Airplane Space Tether Orbital Launch (HASTOL) system. Short of building an actual skyhook, Boeing performed simulations and found that the materials with the necessary tensile strengths—such as Kevlar, Zylon, and ultra-high-molecular-weight polyethylene—can be mass-produced to create cables of the required length and thickness. We have hypersonic planes to reach the cables at the desired altitude and velocity. And the twirling baton, more properly called a momentum exchange tether, wouldn't get pulled down toward Earth after picking up its payload, provided that it is at least two hundred times the mass of the expected payload, hooked at an altitude of

100 kilometers and a velocity of 4 kilometers per second, the study found. Said the authors of the report, "We don't need magic materials like 'Buckminster-Fuller-carbon-nanotubes' to make the space tether facility for a HASTOL system. Existing materials will do."[39]

The basic skyhook or space tether system can be daisy-chained in clever ways to climb higher and deeper into space. One destination in LEO could be a massive orbiting spaceport. With current technology, you could start your trip to the Moon at a spaceport on Earth, fly in a non-rocket spaceplane to a skyhook, and get flung or towed to the orbiting spaceport. From the orbiting spaceport—already moving at about 10 kilometers per second, just like the ISS—you could transfer to a sleek spacecraft for the flight to the Moon. Such a spacecraft would operate entirely in space, never landing on the Earth or Moon and thus not needing the copious amounts of fuel required for leaving these gravity wells. From a spaceport near the Moon, you could descend to the lunar surface. Every element of this system—spaceplanes, skyhooks, spaceports, and spacecraft—can be used repeatedly, much like airplanes and airports, greatly reducing the cost of access to space.

So, why aren't we doing this, if it is technologically feasible? Like so much else in space, there's that pesky catch-22. For this system to be cost-efficient, we would need to make thousands of human trips to space each year. There's currently no such demand.

How much a skyhook infrastructure would cost is a great unknown. Years of research and development are required. The puny ISS, at $100 billion, is perhaps a poor comparison, given the faulty planning that drove the high costs. But $500 billion, similar to the cost of the US Interstate Highway System, is a fair estimate. Once built, the price of moving cargo into space might be reduced from the current $10,000 per pound to $100 or even $10 per pound. A simple, working HASTOL system could reduce the cost of space tourism from tens of millions of dollars to $60,000.[40]

Beyond Rockets: Elevators, Rings, and Towers

Elevator going up—way up. A space elevator is like a skyhook, only the cable reaches down and attaches to the ground. The terminal platform is in deep space. The whole thing is held taut by centrifugal force. Tsiolkovsky, the man behind the rocket equation, developed this concept in the late nineteenth century. Sounds far out, and it is. No cable is strong enough to hold the contraption in place—currently. Let's leave that practicality aside for the moment, though.

A space elevator would provide the cheapest access to space yet, likely only dollars per pound once the system is established. You would merely attach the cargo to the cable and send it up to a platform at geostationary orbit, where the platform remains at the same spot above the Earth, rotating in unison with the land far below it. The elevator ride may take a few days, as you need to climb 22,000 miles, or 36,000 kilometers. That's nearly the distance, coincidentally, of a trip around the equator; and, much like a conventional elevator, you can't climb too fast up the cable without making it unstable. Pray there's no elevator music. The energy needed to climb, though, is trivial. Once you have arrived at the terminal, you are in geostationary orbit, moving at about 3 km / sec, or the rotational speed of Earth itself. From this point, it would take very little energy to travel farther, to the Moon or even Mars in a spacecraft.

Space elevators are just barely within technological feasibility today. We would need 36,000 kilometers worth of cable, and the only known fiber that won't snap under its weight at that length and mass is a carbon nanotube. A carbon nanotube fiber is at least

An artist's impression of a space elevator. The cable is held taut with a counterweight, or terminal, in geostationary orbit 36,000 kilometers above the Earth. Difficult to establish on Earth (primarily for reasons of logistics and terrorism), such an elevator could be built on the Moon and Mars to raise and lower people and cargo.

117 times stronger than steel and thirty times stronger than Kevlar.[41] The longest carbon nanotube spun thus far, though, is a half meter long, a far cry from the 36,000,000 meters needed for just one strand of the space elevator cable. Keep spinning. Should we figure out how to make carbon nanotube fibers cheaply and abundantly, we still have the problem of where to place the cable. The

location would need to be protected from aircraft and—more important and more difficult—from terrorist activity. And should we manage that feat, we next need to worry about the thousands of objects in orbit, collectively known as space junk—mostly discarded satellites. Any of these could collide with the space elevator; there would be absolutely no way to control it. For this reason, despite technological advances, it will be highly unlikely that we ever have a space elevator—on Earth. On the Moon and Mars, however, a space elevator is a very attractive option. There would be, for now, no terrorism concerns. And elevators would be easier to build because the weaker gravity on the Moon and Mars means the cables need not be as strong as they would need to be on Earth.

For Earth, it may be more practical to construct an orbital ring, essentially a maglev (magnetically levitated) railway system around the equator but in low-earth orbit.[42] This would be an engineering marvel, yet it would not require magic materials like carbon nanotubes. Think Saturn's rings, only metal. A metal hoop is built section by section until it reaches completely around the Earth, about 300 kilometers up and 40,000 kilometers long. Set at orbital speed, it would orbit indefinitely, much like the aforementioned chunks of metal called space junk. Running an electrical current through the metal hoop can turn it into a magnet. Now you can hover stationary platforms or pipes over or around this magnetic hoop. It's like a maglev train in reverse; the rails are moving, and the carriage is still. Depending on the strength of the magnet, you could build wide platforms with protective domes where people could live or work. On this platform, you would *not* be in orbit. You would be at orbital height but not orbital velocity. It is the magnetic hoop below your feet that is in orbit, whirling about at 10 km / sec; you, however, are levitating above that on a solid platform, motionless, feeling nearly 100 percent of Earth's gravity. Step off the ring, and you will fall down to Earth (and could feasibly survive in the proper protective gear and parachute).

You could reach the platform via cables lowered to Earth, which help to stabilize the entire structure.

The point of the orbital ring is to provide cheap access to space. Once the ring is built, humans would be able to travel to the orbital platform as easily as they travel 300 kilometers on the ground. You could make a day trip out of it. You could visit for breathtaking views. You might live there as a worker maintaining the orbital ring. Or you might use the platform to propel yourself to deep space. A spacecraft attached to a train-like vehicle, in the absence of wind resistance, could slowly build up the velocity to be flung to the Moon or Mars or beyond. Once at the adequate speed, the craft could detach from the train, fire up its engines, and take off for the stars. You could also use the system to get lots of mass into space, such as solar panels. Humans could collect all the energy we need from solar panels in space, beamed as microwaves to power stations in the ground. The chief obstacle limiting the use of this otherwise unlimited power source is the cost of launching all those solar panels into space.* It would cost many trillions of dollars via rockets powered by chemical fuels. An orbital ring would solve that problem—if it weren't so expensive. The cost of the construction materials—steel, aluminum, and Kevlar—is relatively negligible compared with the cost of launching the estimated 200 million kilograms into orbit. This, too, would cost more than a trillion dollars, a steep price to pay for an untested

*Solar panels in space would be about seven times more efficient than those on Earth because of the near-constant Sun exposure, but beaming the energy down does raise concerns about safety for both people and wildlife.

technology, however theoretically possible it is. Another space catch-22.

Similar to the orbital ring concept is that of a floating high-altitude runway. The runway provides friction and something for the launch vehicle to push against while it accelerates to enormous speeds in the relative dearth of wind resistance or atmosphere. Now, you may have flagged the "floating" part of this description as a showstopper. Runways tend not to float. But you could make them float with "active support." Most structures are maintained through passive support, with beams, trusses, arches, and the like. Active support implies a constant force pushing against an object, such as a thin paper lifted by a stream of air. A bridge, in theory, could be supported by lighter-than-air balloons. Similarly, a rail runway could be lifted and then held aloft by a series of high-altitude balloons—as high as 50 kilometers, which is short of the 100-kilometer space demarcation, or Kármán line, but still high enough to make accelerating to orbital velocities easier. A spaceplane could fly to this runway and transfer passengers or cargo to a waiting spacecraft. Or, cargo could float up with the runway and launch once the proper altitude is reached. To make this approach practical, however, we need to find ways to get balloons higher and then keep them in place, a concept called station-keeping. In August 2018, a NASA team smashed the balloon altitude record by 8 kilometers when they flew an ultrathin balloon to an altitude of 48.5 kilometers above New Mexico.[43]

There are many other rocket-alternative ideas, too, all feasible yet all requiring major investment. One limitation may be the upkeep. Bridges do crumble, and tumbling space infrastructure would be menacing indeed. But spacefaring likely cannot be done via rocket launches for the simple reason of the noise. Even if rockets get cheaper, they won't get quieter. To build a sustainable space economy, we need to move millions of people and cargo and building materials into space, requiring thousands of launches each day. A staggering 100,000 airline flights carry millions of passengers around the world, every day. That kind of traffic going in and out of space in rockets would produce a mind-numbing, Earth-dumbing din.

My prediction: Rocket flights get cheaper by the mid-2020s, driving demand to space; the first space hotels open by 2025, with the first space-based music video and movie segments filmed shortly thereafter; spaceplanes come to fruition, with several companies offering weekly flights to space hotels by 2030 or hour-long flights to opposite ends of the globe; launch costs lower to a point that manufacturing unique commercial goods in the zero-gravity and vacuum environment of low-earth orbit can be profitable; space is a hot destination for the wealthy in the 2030s; launch and descent noise becomes an issue in the 2030s but no solution is put forward, aside from limiting spaceports to remote areas such as deep-sea ports; several large orbiting shipyards and distribution hubs are established near Earth and near the Moon by 2050; first large orbiting space resort with artificial gravity and permanent occupation built by 2050; decades in the planning and making, the first orbital ring becomes operational by the early twenty-second century; large orbiting cities are built by the mid-twenty-second century, many as retirement communities.

4

Living on the Moon

The United States famously landed two astronauts on the Moon on July 20, 1969. Ten more US astronauts followed over the next three years. Some rode buggies; some collected rocks. One played golf, although he never did find the green. And then that was that. On December 14, 1972, astronauts Harrison Schmitt and Eugene Cernan left the Moon after three days on the lunar surface. No human has returned since.

The reasons for our absence from the Moon are many. The primary reason, as noted previously, is that there has been no compelling reason to be on the Moon, given the expense. The purpose for going in the first place, of course, was that the United States was at war with the Soviet Union—a war of political philosophies as well as flesh-and-blood battles fought around the globe by proxy armies starting in the late 1940s. The Soviets threw down the gauntlet when they placed a satellite, Sputnik 1, into an elliptical low-earth orbit on October 4, 1957.

That singular event, taking the United States by surprise, led to the so-called Sputnik Crisis, the widespread anxiety among Western nations that Communist nations had superior technology and thus a superior creed.

One needs to appreciate this anxiety as the sole driver to the Moon. Prior to Sputnik, there was no race, no urgency, no deadline for humans to place themselves beyond the confines of Earth, let alone on the Moon. Quite the opposite—there was a sense of emerging global cooperation to ease the world into the space age. The international scientific organization known as CSAGI (Comité Speciale de l'Année Geophysique Internationale) had held a conference in Rome in October 1954 calling for the launch of artificial satellites during the proposed International Geophysical Year, a period of focus on space technology that would last from July 1, 1957, to December 31, 1958, coinciding with predicted peak solar activity.[1] The intent of these satellites was to map the Earth's surface, and both the United States and the Soviet Union voiced plans to participate. The United States began working on Project Vanguard, a rocket powerful enough to place a three-pound satellite into orbit. The Eisenhower administration knew the Soviets were developing Sputnik, but most in the West were shocked to witness the Soviets launch something so ambitious so quickly. Sputnik was nearly 200 pounds, transmitted data for twenty-one days, and stayed in orbit for three months, a smashing success. The diminutive Vanguard 1, which Soviet premier Nikita Khrushchev mocked as "the grapefruit satellite," wasn't launched until March 1958.

Sputnik had a profound influence on the American psyche. The hysteria set in motion funding for space endeavors that would far exceed what was spent on the development of the first nuclear bomb. Americans had viewed themselves as technological leaders, the nation of Edison, Ford, and the Manhattan Project. Suddenly the nation saw itself lagging far behind the Soviets, whom many Americans considered godless and warlike.

The Soviets followed the October 1957 launch of Sputnik 1 with the November launch of Sputnik 2, this time 1,120 pounds and carrying a dog named Laika. And Laika or not, the United States still wasn't close to being ready. The White House held a public test launch of the Project Vanguard booster rocket on December 6, 1957, and invited the news media to cover what would surely be a fantastic display to restore public confidence. To everyone's dismay, however, the rocket rose only a few feet above the platform and burst into flames. Embarrassingly enough, it was the recently defeated Nazis who helped the United States save face. Wernher von Braun and his rocket team, comprising almost entirely post–World War II German immigrants who had served under Hitler, were recruited by the US Army to work on a rocket project that Eisenhower had canceled in 1955 in favor of the navy's Vanguard project, and successfully launched a satellite named Explorer 1 on a Jupiter-3 rocket on January 31, 1958.* Public optimism was soon crushed again, though, when just a few days later the second test of the Vanguard launch vehicle failed only four miles into the sky, this time higher up for all to see.

Space Race Scorecard

So, who won that space race? You could argue that humanity did. Without the efforts in the 1960s, however motivated by war, we wouldn't be talking about returning to the Moon today. Moreover, the space race inspired hundreds of thousands of young people

* "'Once the rockets are up, who cares where they come down? That's not my department,' says Wernher von Braun." Lyrics by satirist Tom Lehrer from his song "Wernher von Braun" (1965) reflect some of the uneasiness felt over the US enlistment of von Braun. To wit: "Some have harsh words for this man of renown. / But some think our attitude should be one of gratitude, / Like the widows and cripples in old London town / Who owe their large pensions to Wernher von Braun."

worldwide to study engineering and science. The US Congress, warned by college deans that the nation lagged behind the Soviets in educating its citizenry, quickly passed that National Defense Education Act of 1958, which provided millions of dollars to enhance science, math, and language education at all levels. Many of the people I interviewed for this book spoke of a direct, inspirational connection to the Sputnik era. Countless more in their generation, even if they didn't work for NASA, nonetheless chose professions that enabled them to enhance the quality of life for people worldwide through science.

But this notion of Americans "winning" the space race is more in the minds of Americans than in those of the former Soviets. Americans beat the Soviets to a moonwalk, true. But the Soviets excelled spectacularly. Here's the rundown of some Soviet firsts: the Soviets were first to create an intercontinental ballistic missile and orbital launch vehicle; first to put a satellite into orbit; first to put an animal into orbit; first to launch a spacecraft to escape Earth's gravity (Luna 1); first to accomplish data communication to and from space; first to place an spacecraft in heliocentric orbit; first to reach the Moon with a human-made object (Luna 2); first to image the far side of the Moon (Luna 3); first to send a probe to Venus (Venera 1); first to put a human in orbit; first to put a human in orbit for more than twenty-four hours; first to place two men in space together; first to put a woman into orbit (1963!); first to soft-land on the Moon (Luna 9); first to accomplish an orbital rendezvous and docking in space; first to reach the surface of another planet, Venus (Venera 3); and first to coordinate a crew exchange in space. This was all before the Apollo 11 moon landing.

Among the American firsts during the era of the space race were these: the first significant space-based discovery (Van Allen radiation belts, 1958); the first weather satellite; the first geostationary satellite; and, of course, the first human Moon landing.

The Moon landing was indeed sensational. But after that landing, for NASA, it was right back into catch-up mode. The Soviets were first to have a robotic sample return from the Moon (Luna 16); first to place a robotic rover on the Moon; first to soft-land on Venus (Venera 7); first to create a space station (Salyut 1); and first to soft-land on Mars (Mars 3). Into the 1980s, the Soviets were first to create a permanently manned space station, Mir, which operated from 1986 to 2001. Cosmonauts claim the top slots for time spent in space and longest continuous duration in space; US astronauts don't even come close. And one more rarely discussed fact: since the US space shuttle program was decommissioned in 2011, the *only* way to get astronauts to the International Space Station is via the Russian Soyuz spacecraft.

My intent is not to diminish American accomplishments but rather to highlight Soviet accomplishments and elucidate the fact that our current achievements in space are a result of lessons learned equally from both nations. The Apollo moon landings stand out, however, as a pinnacle of human achievement. At its peak, the Apollo program reportedly employed upward of 400,000 people and required the support of more than 20,000 industrial firms and universities—evidence of its warlike mission.[2] This vast team had to create technologies from scratch to meet the ambitious goal set by Kennedy to place a human on the Moon in just over seven years. Launchpads, mission control, spacesuits, miniaturized computers, and, oh right, the rocket—none of these existed. Could your blood flow in microgravity? Could you land on the Moon without sinking into the regolith? Could you send messages to and from a tiny spacecraft so far from Earth? So many unanswered questions. The awesome outpouring of dedication, intelligence, and creativity mobilized to place humans on the Moon has been unmatched in the fifty years hence. In addition, dozens of US pilots and astronauts risked their lives during the Mercury, Gemini, and Apollo programs, and three of them died.* This is why we so rightly celebrated the fiftieth anniversary of the Moon landing in July 2019 with myriad documentaries, books, and magazine articles detailing the fantastic actualization of a bold vision.

That said, Americans didn't choose to go to the Moon and "do the other things . . . because they are hard," as Kennedy proclaimed. United States chose to go to the Moon because the Soviets challenged the US government in a specific direction, and the United States had no other logical choice than the Moon. Had the Soviets devised a plan to drill to the center of the Earth, Kennedy would have used the same language to call for a "core race" to reach the Earth's center by the end of that decade. And maybe we'd have spectacular spin-offs as a result, such as borers to connect cities and continents underground. The road not taken.

I'd argue that the true champion of the Apollo era was not Kennedy but rather Lyndon B. Johnson, the Senate majority leader in 1957 when Sputnik was launched. He was hosting a barbecue at his ranch in Texas when the news of Sputnik crackled through the radio. He watched the satellite streak across the night sky. Later he recalled, "Now, somehow, in some new way, the sky seemed almost alien. I also remember the profound shock of realizing that it might be possible for another nation to achieve technological superiority over this great country of ours."[3] And he did something about it: he encouraged Eisenhower to create NASA; he encouraged Kennedy to set a goal of reaching the Moon; and he carried through with that goal.

*Women trained for space, too. I refer you to the 2018 documentary film *Mercury 13* for the inspirational story of the thirteen women who proved they were capable of being astronauts. NASA declined to accept them into any official training program.

As for the difficulty of getting to the Moon, even with robots, how telling it is that no one was able to claim the Google Lunar XPRIZE nearly fifty years after Apollo. That privately funded challenge offered a $30 million purse to anyone who could land a rover on the Moon, travel 500 meters, and transmit back high-definition video and images. The purpose was to incentivize space entrepreneurs to create a new era of affordable access to the Moon and beyond. The challenge launched in 2007 and, in 2015, was extended to 2018. But no competitor even came close. Mind you, only nongovernmental entities were allowed to compete. Otherwise, China would have claimed the prize.

Enter China

China has ambitions. From that nation's perspective, it suffered through what they describe as "one hundred years of humiliation," when the country fell from its status as a self-perceived world leader in the early nineteenth century and subsequently was "attacked, bullied, and torn asunder by imperialists."[4] The rise of the Communist Party in the late 1940s and the founding of the People's Republic of China in 1949 was the beginning of the end of humiliation. Now China wants to reclaim its greatness, built on a centuries-old concept of *fuguo qiangbing,* meaning "rich country and strong army."[5] Space and, in particular, the Moon factor prominently in China's desire to be a rich, strong country. Indeed, its space program is an extension of its People's Liberation Army, and China views space as the ultimate high ground from which protection could be secured (and war could be waged).[6]

In December 2013, China landed a rover named *Yutu* that, by October 2015, had set the record for the longest lunar rover operation. Although not widely reported in the United States, China has had several missions that have orbited or landed on the Moon, including a successful landing on the far side in 2019, a first. China has launched several modular space stations; and it has a working rocket fleet known as the Long March 2, which includes the Long March 2F, one of only two rockets that carry humans into orbit (the other being Russia's Soyuz rocket). It bears repeating: China and Russia can launch humans into space; the United States, Europe, and Japan cannot.

The China National Space Administration (CNSA) has announced plans to send a probe to drill several meters into the lunar surface and return samples to Earth by the early 2020s. CNSA also announced plans to send astronauts, or taikonauts, to the Moon by the close of the decade, with the goal of establishing a permanent base on the lunar south pole. And the long march to that goal is in motion.

In May 2018, China completed a yearlong test of its Yuegong-1, or Lunar Palace 1, a simulated, nearly self-sufficient lunar habitat with room for four. The experiment, conducted at Beihang University in Beijing, engaged eight volunteers who rotated through the habitat for several months at a time. This was a test not of human endurance but rather of a closed system with plants. The habitat recycled 100 percent of its air and water, and the volunteers grew 80 percent of their food calories, according to Chinese news reports.[7] The volunteers grew an impressive array of food, such as wheat, peanuts, lentils, and fifteen different vegetables. One novelty was the addition of mealworms as a protein source; the worms lived off discarded vegetable matter. The habitat comprised two farming areas about sixty square meters each and a living area about forty square meters with three bedrooms, eating area, and bathroom.

It seems possible that China will be on the Moon with humans before NASA or ESA. The Chinese government released an eight-minute video in conjunction with the Yuegong-1 habitat, describing life on the Moon and beyond. It opens with children gazing into space, dreaming of visiting the stars. Then a powerful rocket launch consumes the screen. It was just like one of those high-quality NASA videos, only everyone was Chinese and speaking Chinese. You see, in China's version of *Star Trek,* the lingua franca of the galaxy would be Mandarin; Captain Kirk would a six-foot, brown-eyed, black-haired Han; and any token American surely would be relegated to the role of a yeoman.

To Jack Schmitt, the next to last man to walk on the Moon, China's ambitions present a "geopolitical imperative" for a US return to the Moon. The situation right now is very comparable to the one that we faced in 1960 with the Soviets, he said. "If you're not seeing that, you're not paying attention," he told me.[8] The concept of a new space race already has permeated US military and political circles. This may sound ominous, but it may be good news for space enthusiasts eager to create settlements on other words. Schmitt is among them. He wants to see the United States set up mining operations on the Moon, primarily to mine helium-3, a potential fusion fuel. Without China eyeing the Moon for the same reasons—mining rights—the United States might have only a passing interest in returning. But the tide has turned. It just doesn't have a plan that remains stable through each change of administration.

In May 2019, NASA announced a White House directive to place an American man and woman on the Moon by 2024. Called Project Artemis, the plan compresses to five years the previous, overly ambitious Trump administration call for a 2028 human return, announced just two months earlier. To the outsider, this may sound like a bold and decisive move to return the United States to space glory. But, in reality, this critically flawed move is bereft of lessons learned from previous US administrations—and a recipe for failure. If the goal is to establish a permanent lunar presence, rushing humans there by 2024 only adds complexity and cost with little gain. The plan is impulsive, in contrast to China's methodical approach.

The United States should first send probes and robots to establish infrastructure for humans to utilize when they finally arrive. That *was* NASA's and ESA's plan until the White House forced NASA to shuffle priorities, cancel programs, and reimagine projects currently in development.

What will this Project Artemis space couple do on the Moon? It hasn't been announced, but they surely won't be building shelters, securing water, making breathable oxygen, creating rocket fuel, and laying down tracks for lunar surface transport, as robotic devices would do. To wit, if they make it, they will have gone and returned and spent $30 billion without advancing the cause of a permanent return to the Moon. This ill-fated lunar mission will almost certainly be canceled by the next administration in 2020 if Donald Trump is not reelected president. If it does survive 2020, the mission will make the 2024 deadline only with a major infusion of extra cash. The White House offered only one year of guaranteed funding, at $1.6 billion. Additional funding would need to come from the US Congress, who may not favor this plan for reasons both partisan and fiscal. According to the US Government Accountability Office (GAO), in a report issued a month after the announcement of the 2024 return to the Moon, NASA has fudged the true cost of the Space Launch System (SLS), the rocket meant to take astronauts and materials to the Moon and beyond. SLS is years behind schedule and billions of dollars over budget. The GAO found that NASA and Boeing, the SLS's primary contractor, "underestimated the complexity of manufacturing and assembling the core engine stage section."[9] The first

SLS launch, once scheduled for 2017, then 2019, then June 2020, will slip to at least June 2021.[10] The SLS first needs to make several launches to get the LOP-G space station in space before taking humans.

Whatever the ultimate plan might be for the United States to return to the Moon, it won't include a collaboration with China. In 2011, the US Congress banned NASA from working closely with China. As stated under Public Law 112-55, SEC. 539, "None of the funds made available by this Act may be used for the National Aeronautics and Space Administration (NASA) or the Office of Science and Technology Policy to develop, design, plan, promulgate, implement, or execute a bilateral policy, program, order, or contract of any kind to participate, collaborate, or coordinate bilaterally in any way with China."

This so-called Chinese exclusion policy is not without cause. A 1998 congressional investigation, now known as the Cox Report, determined that technical information provided to China by American aerospace companies concerning commercial satellites ultimately improved Chinese intercontinental ballistic missiles technology, China's primary motive. This led to an almost immediate embargo on information-sharing with China, culminating with the 2011 act signed by President Obama. Relations had grown worse after China, in 2007, intentionally destroyed its own weather satellite during a test of its antisatellite missile technology—proving it could destroy any US satellite while simultaneously creating the largest collection of orbital debris in the history of human endeavors in space, as estimated 150,000 high-speed particles, making low-earth orbit more dangerous for everyone.[11] (Truth be told, the United States and Russia blew up a satellite decades ago, and India did this in 2019, albeit at a smaller scale. Unfortunately, this is how some nations demonstrate that they can be just as destructive as the next guy.)

Moon Rush

The Moon's surface area is about 14.5 million square miles, or 38 million square kilometers. Although that's only about 8 percent that of the Earth, it is still about the size of Asia. Call it the eighth continent. No one owns the Moon, and according to the Outer Space Treaty—formally known as the Treaty on Principles Governing the Activities of States in the Exploration and Use of Outer Space, including the Moon and Other Celestial Bodies—no one is allowed to own it. More specifically, the treaty states that "outer space, including the moon and other celestial bodies, is not subject to national appropriation by claim of sovereignty, by means of use or occupation, or by any other means." More than a hundred countries have signed this treaty, including China and the United States.

What the Outer Space Treaty means in terms of settlements on the Moon is anyone's guess, though. Clearly this has never been challenged. Many lawyers believe that, although nations cannot claim territory, they can claim and profit from resources, such as minerals. Indeed, this notion scares some smaller, non-spacefaring nations that see themselves being left out. In 1979, a group of nations crafted the Moon Treaty, formally known as the Agreement Governing the Activities of States on the Moon and Other Celestial Bodies. This treaty holds that lunar resources belong to all of humankind and cannot be exploited for private, commercial, or national gain. The Moon Treaty has been ratified by only eighteen countries, none of which currently operate in space. In the United States, the L5 Society—folks who wanted to build orbiting space cities, mentioned in Chapter 3—rallied Congress to reject the Moon Treaty on the grounds that it would thwart space development. In short, they argued, what company would invest in lunar infrastructure if there were no possibility of profit?

I believe the likely scenario will have the Moon looking, at first, a lot like Antarctica. Seven countries claim territory in Antarctica, although none are recognized by the broader international community. There's an Antarctic Treaty that does not allow rights of sovereignty on the continent, bans military activity, and prohibits mining at least until the year 2048 (although some believe China's aggressive mineral explorations under the guise of science doesn't bode well for preserving Antarctica's pristine environment).[12] Antarctica's resources, minerals, and fuels are tucked away below miles of ice in a formidable climate and isolated location, bereft of infrastructure for export, and thousands of miles from the nearest market. It's just not that profitable to mine Antarctica unless the ice starts to melt and mineral and fuel resources elsewhere in the world are exhausted.

As in Antarctica, as access to the Moon becomes relatively affordable, it will first be a location for conducting scientific experiments, with a dedicated crew maintaining bases for a few years at a time as scientists, engineers, and prospectors come and go. If lunar resources prove to be profitable, all bets are off. Countries and major corporations will scramble to claim the best "land" and argue their case vis-à-vis the Outer Space Treaty in the International Court of Justice. When the rubber hits the regolith and when trillions of dollars in profits are realized, treaties will be redrawn or otherwise reinterpreted to allow for the commercialization of lunar activities. The history of colonization, after all, is the history of the exploitation of resources—not shared wealth among nations but rather national appropriation, duplicity, treaty abandonment, and war.

One space exploration company, the Shackleton Energy Company, has long-standing, solid plans to mine the water on the Moon and transport it to a resupply station in low-earth orbit. Water and its constituent oxygen and hydrogen would be used to supply spacecraft with water to drink, air to breathe, and hydrogen to burn. Shackleton Energy is led by the charismatic Bill Stone, who has set records in cave exploring, plumbing abysses thousands of meters deep. He has worked with NASA in developing a prototype of a mission to the moon Europa, with a probe that can bore through the thick ice into the liquid ocean there. And he wants to personally lead what he calls an industrial "Lewis and Clark expedition" to Shackleton Crater to create his own fuel to get back off the Moon. Shackleton Energy likely has the right to use lunar water, but can the company sell it? No one really knows. There is a market value established already, though. Having fuel on the Moon could reduce the cost of going there by a third, because the rocket leaving Earth wouldn't have to carry the fuel for the return trip. United Launch Alliance has stated that it would be willing to pay $500 / kg for propellant on the lunar surface and $1,000 / kg in orbit, and the company would need at least 1,000 metric tons; NASA has estimated a need for 100 metric tons of propellant for ascent vehicles.[13] Ice mining could foster a billion-dollar industry and could serve as the biggest money-maker on the Moon in the short-term.

Depending on concentrations of resources and the price of excavation and of transport—three gaping unknowns that have kept this discussion purely in the realm of speculation for decades—lunar resources can be profitable for three different markets: here on Earth, on the Moon itself, and for space exploration.

Here on Earth we could possibly benefit from two kinds of lunar resources: helium and rare-earth minerals. Helium-3, an isotope of helium that's rare on Earth but comparatively abundant on the Moon, is an ideal fusion fuel. If we can commercialize fusion with helium-3, we will have plentiful, clean "green" energy. It wouldn't be cheap, because it's coming from the Moon, after all. But it could be comparable in price to fossil fuels per kilowatt of energy. The rare-earth minerals are composed of rare-earth elements, from that region of the periodic table set off from the rest—elements such as cerium (Ce), gadolinium (Gd), and yttrium (Y). These elements and minerals are crucial components of modern gadgetry. Dysprosium and terbium are used in touchscreens to produce color; gadolinium enhances MRI images. Contrary to their name, they are quite common. The problem is that they are harder to extract from the ore containing them than are less common elements such as gold and copper. Extraction may be easier and safer on the Moon. KREEP, a conglomeration of potassium (K), rare-earth elements, and phosphorus (P), found in abundance on the Moon could be shot all the way back to Earth and dropped into the ocean for pickup. Potassium and phosphorus would certainly be welcomed on Earth for soil fertility.

To live on the Moon and mine these kinds of minerals, the most important resource will be water, which is in the form of deeply frozen ice. Again, that's water to drink, oxygen to breath, and hydrogen to burn—all translating to huge cost reductions for lunar activity if this can be extracted locally. There appears to be more water than we thought. Indeed, the perceived lack of water was one reason why so few nations were interested in establishing lunar bases through the 1970s and 1980s. The discovery of lunar ice at the poles as recently as the 1990s by NASA satellites rekindled interest in the Moon as a possible base location. In 2009, the Indian Chandrayaan-1 mission detected widespread presence of water molecules across the lunar surface. A 2010 analysis found that water deposits in shadowed craters might be as concentrated as 8 percent by mass, mixed with regolith; a 2018 analysis upped that to 30 percent in parts.[14]

Suddenly, as a result of these discoveries, the Moon appears to be that much more hospitable. The Moon also has all the industrial elements needed to build structures there and nearby, including iron, silicon, aluminum, magnesium, titanium, chromium, calcium, and sodium. The Moon has uranium, too. None of these elements are needed badly on Earth, but they are needed on the Moon and in space for spacecraft and orbiting cities. Raw materials could be flung off the surface of the Moon with relative ease because of the low gravity.

One more lunar resource would be intellectual: there is considerable scientific knowledge to be mined from the Moon. The leading theory for the origin of the Moon is that it was created from the collision of a Mars-sized protoplanet with Earth 4.51 billion years ago, only about 30 million years after the Earth itself was formed. Studying the "lunology" of the Moon will reveal what the early Earth was like, a geological record that has been lost because of the Earth's surface constantly reshaping over the eons. The Moon also would be an excellent platform for many types of astronomy and for practicing how to send humans deeper into space.

A Challenging Lunar Industry

Depending on cost of accessibility and concentrations of profitable resources, the Moon could become a science and industrial park with some tourism in the next twenty years. Given the Moon's proximity and promise, it is inevitable that we set up shop there at some point. Costs are falling, and data are providing ever more optimism to speculators about resource concentrations. Interestingly, because of the low lunar gravity, it is more energy efficient to ship materials from the Moon a full 250,000 miles back to low-earth orbit, where the ISS is, than it is to launch them two hundred miles up from Earth's gravity well. Massive spacecraft could be built with lunar materials in shipyards in lunar or Earth orbit, with only the lighter, high-tech parts coming from Earth.

Mining operations in wealthy countries are becoming increasingly automated, done by robots controlled remotely by humans. Such would be the case on the Moon. Already, numerous nations are prospecting the lunar surface via satellite using remote-sensing technology to find out which deposits are where and in what quantities. Miniaturized, lighter-weight excavators are being built to be transported to the Moon. Chemical plants used to extract minerals and various volatiles are being miniaturized to process materials in situ. The low gravity on the Moon means that excavation is, in some ways, simpler than it is on Earth. The Moon exerts only one-sixth the gravitational force that Earth does. Stuff comes up more easily. Support structures can handle more mass because the weight of material, a function of gravity, is lighter. And because there is some gravity, unlike on the ISS, ores and minerals of various densities will sink or rise as they do on Earth, so no special machines are needed to compensate for weird microgravity physics. Daunting disadvantages include the fact that drilling can cause too much heat from friction, with no air to cool it. Also, blasting is extremely dangerous because projectiles would travel faster and farther then they would on Earth.

But one great advantage of a low-gravity environment is the ability to sling the raw materials off the surface in a device called a mass driver. The Moon's escape velocity is about 2.38 km / s, which is only about a fifth of the escape velocity of Earth, 11.2 km / s (25,000 miles per hour). This means you don't need a rocket to blast off from the Moon. All you need to do is propel an object down a track faster than 2.38 km / s and it will leave the surface of the Moon instead of rolling over the horizon. Now, 2.38 km / s is fast—about twice as fast as a bullet fired from a gun. But this velocity is attainable on the Moon with a maglev rail system, aided greatly by the complete lack of air resistance. A mass driver is like a slingshot on such a rail. Cargo could be accelerated for several kilometers until it reaches sufficient speed and then let go. In this way, workers on the Moon could load raw materials into a mass driver and fling them to some desired point in orbit, where they can be captured and used to build spacecraft or solar panels or other massive orbital objects. Sure, the devil's in the details, but this does not require advanced physics or engineering.

Much of the mined lunar materials would be for the Moon or orbital space activities. For example, workers could mine for rocket fuel components—hydrogen, oxygen, aluminum, or magnesium—and launch or fling them to a fueling station in orbit for spacecraft to use. Rocket fuel derived from water, pumped into spacecraft at orbiting filling stations, offer tremendous savings to spacecraft owners because developers could build smaller craft with no need to accommodate all the fuel required for the full trip.[15] If uranium could be enriched on the Moon, we could use that for nuclear fuel to power factories or habitats or nuclear-powered rocket ships. Unlike on Earth, there would be little concern about nuclear accidents because all life on the Moon would be confined to habitats and environments protected by airlocks.

Earlier I mentioned the catch-22 that pops up in space discussions: the industrialization of the Moon makes sense if and only if there's a space and Earth industry needing the materials. Living and working on the Moon is feasible, albeit challenging. But the challenge won't be attempted unless there's a profit to be had from the industrialization. First, let's examine helium-3, or ^{3}He, as a fusion fuel.

Helium-3

Fusion is the power source of stars. The Sun fuses lighter elements into heavier ones, primarily hydrogen into helium nuclei and helium into carbon. In the process of two atoms being squeezed into one, some mass is squeezed out. And mass, as expressed in Einstein's famous equation $E=mc^2$, is equated to energy—lots of energy. The Sun can sustain fusion because of the high temperature and pressure in the core, all powered by crushing gravity. We mortals cannot create fusion without using more energy than we get out of it. We can pull off *fission,* the splitting of atoms, but not sustained fusion. Hydrogen is very hard to fuse; like repulses like. Fusing isotopes of hydrogen, called deuterium (^2H) and tritium (^3H), is a little easier, but this produces high-speed neutrons, which are difficult to contain and which make other materials radioactive upon collision. Hydrogen contains a proton and electron but no neutron; deuterium and tritium contain one and two neutrons, respectively. The hot, dangerous neutrons are by-products. But ^3He is missing a neutron. Fuse that with deuterium, and you end up with regular helium (^4He) and a proton, which is just a hydrogen nucleus. The equation is $D + {}^3He \rightarrow {}^4He + p + 18.4MeV$ energy. Similarly, $^3He + {}^3He$ would yield ^4He and two protons and a similar amount of energy. Either way, protons are safer to handle than neutrons, and there would be no radioactivity. Thus, ^3He-based fusion is the cleanest, most powerful energy within our reach. Just 100 kilograms of ^3He, a little more than the weight of an adult male, could power a city for a year.[16] Try that with oil.

The Earth has very little ^3He, only a few parts per trillion in the atmosphere; but the Moon is coated in the stuff, relatively speaking, estimated between 20 and 30 parts per billion near the equator. That's more than a million tons in the first few meters of regolith, at a potential energy sales cost of billions of dollars per ton. The ^3He has been deposited on the lunar surface for eons by solar winds. Harrison Schmitt, a geologist who gathered and subsequently analyzed lunar rocks from the Apollo 17 mission in 1972, estimates that we could mine 100 kilograms of ^3He from two square kilometers of lunar regolith to a depth of three meters. In Schmitt's mind, ^{3}He is the only fuel that can meet the world's growing energy demands, given the climate change issues surrounding the burning of fossil fuels, the radioactive dangers of nuclear fission, and the relative weakness of terrestrial solar and wind power. Going all the way to the Moon for ^{3}He would incur a cost, but the bang for the buck would make ^{3}He comparable in price per watt to coal, without the pollution, Schmitt says. That same regolith contains titanium and other valuable elements, so by-products would be profitable. Regular helium, ^{4}He, is measured in parts per million on the Moon, a thousand times more common than ^{3}He, and is increasingly in short supply on Earth. So that's one more potential lunar export.[17]

Schmitt's 335-page book called *Return to the Moon* is based entirely on this premise of mining ^{3}He, explaining the minutia of the mining process and the profits to be had. The catch? Fusion using ^{3}He isn't a thing yet. And it is far from guaranteed that having lunar ^{3}He here on Earth will make it happen; ^{3}He fusion is more difficult to achieve than deuterium and tritium fusion. So, we'd have to accomplish that kind of fusion first and then master ^{3}He fusion with our Earth supply before we set off to mine it on the Moon. Schmitt thinks we're very close to commercial fusion with deuterium and tritium, that is, where power out is greater than power in. He thinks it is merely a matter of placing more money into research. Many nuclear engineers, however, say we are not so close. Not mastering ^{3}He fusion would be a major blow to the profitability of returning to the Moon. With ^{3}He fusion, then the Moon is more profitable than all of OPEC. Without it, we're down to lunar water and rare-earth elements to drive profits.

The other limitation is that most of the ^{3}He is at the equator, where it is harder to mine because of the dramatic temperature fluctuation. Concentrations are low at the poles. And like a fossil fuel, once the ^{3}He is mined and used, it's gone for good. How long would resources hold out? A generation? A century? Is the investment worth it for short-term gain? Is there a commercial market, or will solar be more competitive? Depending on investment costs and mining efficiency, it may be more economically feasible to beam solar energy to Earth from solar panels on the lunar equator or in orbit, or to invest in some other clean, renewable energy method.[18] In this regard, many scientists view the prospect of lunar ^{3}He mining to be, well, lunacy.

Rare Earths

Next, we have the lunar supply of rare-earth elements, which is of potential high value. As noted previously, these elements and minerals aren't necessarily rare on Earth but just hard to extract from the ores containing them. This is nasty, polluting work—so nasty that the United States has stopped mining rare earths and prefers to import them from China, which, frankly speaking, at this moment in its development, values quick economic development over long-term environmental health. As of 2017, China was mining 80 percent of the world's rare-earth elements, some 105,000 metric tons. Australia was a far second, with 20,000 metric tons, followed by Russia (3,000 MT), Brazil (2,000 MT) and India (1,500 MT).[19] Worrisome for many countries is the fact that China appears to be hoarding its supply to drive up prices.

Mining rare-earth elements on the Moon would shift the environmental concerns to a dead rock, which many would argue is more desirable, but here's the catch: China can play the hoarding game for only so long, until other countries decide to increase their production. The demand is so high—an iPhone uses all seventeen rare-earth elements—that countries will either bear with the environmental cost or else find ways to more safely mine or recycle the elements. Unlike with ^{3}He, the price of rare-earth elements would need to skyrocket before any country turns to the Moon as a source. Also, other forms of mining would need to be established on the Moon first for rare-earth mining to be economically feasible.

Iron and the More Mundane

As exotic and promising as ^{3}He and rare earths may sound, the future of lunar mining may lie in the more mundane excavation of ice, iron, aluminum, and the like. The reason is twofold. First, these basic building blocks of industry can be used to construct a space infrastructure. Unless humans invest in a launch infrastructure that can lift construction materials into space for pennies a pound, it will not make sense to use Earth materials for anything used or built in space. A vibrant space infrastructure of orbiting cities, solar panels, lunar settlements, mining, shipbuilding, and Solar System exploration will utilize mostly lunar materials (and eventually asteroid materials). In the same way that colonists in the New World did not bring wood from the Old World to build their homes, a spacefaring human race will use materials that are close at hand. The Earth is simply in too deep of a gravitational well to supply the materials needed for space.

Compared to asteroids, which I discuss in Chapter 5, the Moon has higher concentrations of aluminum, titanium, and uranium. Titanium is in the form of the mineral ilmenite, $FeTiO_3$; mining it would have iron and oxygen as by-products, both of high value on the Moon. Processing this will be challenging but might be done with concentrated solar ovens or microwaves.

Uranium could fuel a nuclear reactor. Aluminum could be forged into support structures for habitats or solar panels. At its crudest, the first mining could be as simple as scooping up regolith and heating it a few hundred degrees to collect any off-gassing volatiles (hydrogen, helium, carbon, nitrogen, fluorine, chlorine). Dig sites could be anywhere, with no concern for damaging a living environment, and nothing dug up would go to waste. Angel Abbud-Madrid, director of the Center for Space Resources at the Colorado School of Mines, says there already has been sufficient remote prospecting of the lunar surface to warrant sending rovers and other machines to the Moon now to test the important technology of robotically extracting and processing resources.[20]

The lunar regolith is more than 20 percent silicon by weight, which could be used in combination with aluminum and other elements to make solar cells for energy generation on the lunar surface or in orbit, to beam to Earth.[21] Enough solar panels placed in Earth orbit, with energy beamed as microwaves to collectors on Earth, could provide for all our day-to-day domestic energy needs and many industrial needs. We know solar panels work in space because they power most satellites. There's never a cloudy day in space. This concept of orbiting solar panels looked attractive in the 1970s, when energy prices were rising and when US president Jimmy Carter put thirty-two solar-thermal panels on the roof of the White House to heat water as an example for the nation to follow. Then oil prices fell, and US president Ronald Reagan removed these working solar panels as a different example for the nation to follow.[22] Solar energy harvested by orbiting solar panels remains a reliable, renewable, clean energy source, nonetheless. The primary hindrance is the cost of the infrastructure, and that's determined by the cost of access to space.[*] If launch costs come down, the Moon could become an energy boomtown.

The second reason for mining the Moon is that Earth's resources are largely finite. We'll set the environmental discussion aside for the moment. At some point, metals and other industrial resources will be too expensive to extract as a result of scarcity, and it becomes an economic argument. Digging deeper and deeper incurs a higher cost. We could reach a point at which lunar materials may be cheaper to extract. As for the environment, there's little debate that mining for resources causes irreparable damage to the Earth in terms of contamination of groundwater, surface water, and soil by chemicals from mining processes, as well as erosion, creation of sinkholes, and the inevitable loss of biodiversity. The world's largest polluters would concede to this much. The argument actually is the trade-off between environmental damage and economic development. Many argue that the economic development is worth it—and who in the developed world can pass judgement, living in countries with a history of denuding landscapes.

The already-lifeless Moon may very well become a practical, vast, open mining pit for the benefit of life on Earth, a concept that admittedly will generate LUCA.[†] But humans may come to tolerate the "spoiling" of pristine lunar regolith because it is so remote and inherently barren. The Moon's far side isn't even visible from Earth. Sadly, the worst-polluting mines on Earth are the ones too remote to be seen, in impoverished countries, employing or enslaving children. The horrific mining operations of the materials to make your smartphone are out of sight and out of mind.

[*] A secondary hindrance is safely beaming energy to collectors, a plausible but untested technology on a commercial scale.

[†] LUCA: loud, unfriendly counter-argument.

On the Moon, there will be no child labor and no reduction in biodiversity. Moreover, as poorer countries become wealthier, they may place greater restrictions on mining and drilling, as we saw decades ago in the United States, Europe, and Japan, and more recently in Brazil. This may force us to mine the Moon.

In short, basic lunar materials can help the human race. One direction we can take—and there's nothing wrong with this, if it is possible—is to forget about space, reduce the world's population, become more efficient with finite resources, strive to lift all humans out of poverty, and provide all humans with access to the technology, education, and mobility currently realized in developed nations, all the while not extracting any more resources (wealth) from the Earth. Or we can tap into lunar and other space resources as new sources of wealth, all the while making Earth more livable.

Science on the Moon

Science on the Moon will bear an uncanny similarity to that performed on Antarctica: unique geology / lunology and astronomy. NASA has a list of 181 science objectives for the Moon. This is mostly the study of the Moon, Earth, asteroids, and comets via rocks on the lunar surface and observations of the Earth, Sun, and beyond. Among bold but feasible ideas are massive telescopes on the far side of the Moon. That's the side of the Moon that always faces away from Earth. There is no dark side of the Moon; that's a misconception. The far side gets the same amount of sunshine as the visible side. (It's completely dark only during a full moon, when the side we see is lit.) Nevertheless, the far side is the perfect location for radio telescopes because there would be no interference from Earth. We could build telescopes in other wavelengths, as well. The lack of atmosphere means seeing is always perfect; no wavelengths are blocked from reaching the lunar surface. Infrared telescopes would work particularly well in the frigid darkness of a crater. Also, because of the lower gravity, dishes could be larger than they are on Earth while remaining structurally sound. In this regard, the Moon is the ultimate mountaintop.

Another kind of lunar science will come from the very act of living there. We can study the health effects of one-sixth Earth gravity. As discussed earlier, we currently have only two data points: 1G (on Earth) and 0G (in orbit). How bodies respond to 0.16G on the Moon may provide some insight on the chances of long-term survival at 0.38G on Mars. If bone and muscle loss is 16 percent improved on the Moon compared with the ISS, then this could imply the health effects of low gravity follow a linear path between 0 and 1. If it's much better than that, this bodes very well for 0.38G indeed. Similarly, everything we do on the Moon—building shelters, growing food, extracting water and other resources, wandering about—will be practice for living farther from Earth. Mars is six to nine months away, which is a long time to wait for help, should emergency strike while living there. The Moon is only three days away. We could send materials there even faster, in less than a day, at the price of using more fuel. The Moon is an undeniably safer place than Mars to "practice" being a spacefaring species.

Where to Set Up Camp

China's space ambitions have nudged all the major space players—the United States, the European Union, Russia, Japan, and emerging India—to speak of not merely returning to the Moon but going back for good. This is not hyperbole. Rather, this is a reflection of the fact that journeying to the Moon "Apollo style" just to plant a flag and collect a few rocks serves no purpose in this day and age.

Both technology and our understanding of the Moon has advanced to such a degree that many think Moon bases are feasible and relatively affordable, if not profitable in the long run.

Table 4.1, comparing the Moon to Mars and the lunar poles to the lunar equator as settlement areas, shows that there's no ideal place to set up camp. There are trade-offs. I introduce Mars here because some argue that we should forget about the Moon and go straight to Mars; others see the Moon as a necessary stepping-stone to Mars. I think arguments on both sides are flawed. Going from the Moon to Mars is not going from hard to harder but rather from hard to similarly hard. Moreover, a stepping-stone implies a transition to something bigger and better, and I don't see Mars as superior to the Moon; they would serve different purposes in our expansion into the Solar System. I also don't see us leaving the Moon behind once we are on Mars, as a stepping-stone implies.

The longest stay on the Moon was in 1972, when Eugene Cernan and Harrison Schmitt spent seventy-five hours on the lunar surface. This was the most science-rich of the Apollo missions and included extensive geological sampling (Schmitt was a geologist and the first scientist in space) as well as biological studies on five mice that accompanied the astronauts (named Fe, Fi, Fo, Fum, and Phooey, the first rodentianauts on the Moon). Living on the Moon for more than a few days, however, will be challenging. Cernan and Schmitt spent a total of twenty-two hours exposed on the surface, cruising about in their rover, and the rest of the time in the relative safety of their lander.

A Moon base would need to offer a multitude of protections.

The foremost and ever-present threat is solar and cosmic radiation, as discussed in Chapter 2. The Moon has just a whisper of an atmosphere, about ten trillion times thinner than Earth's. This is essentially equivalent to the vacuum of space. The negli-

Table 4.1. What's the best place for a settlement? There are pros and cons when comparing the Moon to Mars, but also when comparing the Moon to itself.

The Moon versus Mars

Moon	Mars
Pros	**Cons**
accessible in three days	takes six to nine months to reach
near-instant communication to Earth	communication to Earth lags, eight to forty minutes
Earth is a beautiful, familiar figure in the sky	Earth is seen as a lonely point of light
minable minerals; easier to process and export	minable minerals; harder to process and export
low gravity good for coming/going and flinging raw materials into space	higher gravity means coming/going is challenging
Cons	**Pros**
low gravity likely not suitable for raising children	gravity may be enough to raise children
dearth of volatiles (nitrogen, carbon)	plenty of volatiles; everything needed for life
nights last about two weeks	day/night cycle nearly identical to Earth
extreme day/night temperature swings	workable day/night temperature swings
no atmosphere	CO_2 atmosphere, albeit thin
toxic dust clings to everything	greater variety of minerals

The Lunar Poles versus the Lunar Equator

Lunar Poles	Lunar Equator
Pros	**Cons**
steady temperature ~−50°C	day/night temperature fluctuations −173°C to 127°C
crater rims with 85%–100% sunshine	dark of night lasts fourteen days
abundant ice in craters	dearth of water
Cons	**Pros**
fewer minerals of industrial interest	plenty of helium-3 and other minerals of interest
solar wind creates electrical charge on crater rims	easier to fling material into equatorial orbit

gible atmosphere and near absence of a magnetosphere means that deadly radiation bombards the lunar surface all day and all night. Had there been a serious solar flare when Cernan and Schmitt were on the Moon, they could have died from radiation poisoning. So any Moon base, regardless of the location, would have to be underground or otherwise covered with more than two meters of lunar soil, regolith, to block the radiation from entering the body.[23]

Nearly every lunar settlement scenario conceived since the 1950s envisions interconnected domes dotting the lunar surface. This idea remains feasible as long as the dome walls are thick enough. ESA has proposed sending robots to the Moon to create igloo-like structures, inflatable habitats covered in a two-meter-thick layer of regolith.* Because of the low gravity, the structure doesn't need to be that sturdy to support this mass.† Instead, the mass is good for maintaining the integrity of the inflatable dome. Remember, there's no air pressure on the Moon, so an inflatable will expand and can pop, just like a balloon in the thin air of the Earth's stratosphere does. In this ESA concept, an unmanned lunar lander touches down on the Moon and detaches a habitation capsule, a cylinder a few meters wide and about twice as long. This capsule then deploys two robotic devices on wheels from one side of its hatch and a large, inflatable shelter from the other. The robots—controllable from Earth or from a proposed space station in lunar orbit—proceed to scrape up regolith and, over the course of three months, coat the inflatable structure through a process similar to 3-D printing. The capsule serves as the airlock to the

*This depth is suitable only for short-term stay; more than five meters of regolith is needed for complete protection.

†I shouldn't be too cavalier; moonquakes are not uncommon, particularly thermal quakes caused each morning when the Sun warms the lunar crust.

shelter. The pressurized habitat, possibly adorned with cupolas to allow filtered sunlight to enter, is suitable for four humans, who arrive after their prefabricated lunar home is completed. This scheme has been tested successfully on Earth.

Such a dome protects against cosmic rays, solar rays, small meteoroids (a big problem), and temperature fluctuations, and it provides an earthlike mini-environment with suitable air pressure and oxygen—the very basics we would need on almost every Solar System body. Master this for the Moon, and there's no reason you can't do it on Mars or even Pluto with minimal tweaking. Similar shelter schemes entail what appears to be a dome but are, in actuality, a large area mostly underground, with just the top part showing. Such schemes date back decades but require massive excavation and imported materials, highly impractical for the near future. As noted previously, it takes a lot of fuel not just to launch materials from Earth but also to land heavy equipment and supplies on the airless Moon. ESA's plan to rely on local lunar resources greatly reduces the cost and complexity of the project. Moreover, these habitats could be built one by one and connected underground, creating a village slowly over time.⋆

We should be able to make bricks from the regolith, too. Bricks usually require water, a precious commodity on the Moon, but engineers have proposed fabricating water-free bricks by mixing regolith and sulfur or by high-temperature baking of the regolith itself.[24] The former method is simple but produces a weak brick; the latter is energy-intensive. Either way, the bricks

⋆ESA's "Moon Village" concept is often misinterpreted to mean the construction of a village on the Moon. But the "village" ESA envisions is a community of scientists and engineers on Earth coming together to share ideas to make lunar settlements a reality.

An artist's impression of the European Space Agency's proposed Moon base. Once assembled, the inflated domes are covered with a layer of 3-D-printed lunar regolith by robots to help protect the occupants against space radiation and micrometeoroids. Several meters' worth of regolith is needed to fully protect occupants.

would need to be glazed or otherwise sealed to make them airtight. Not trivial.

That takes care of radiation. But days are scorching hot and nights are cold on the Moon. Because there is no atmosphere to trap and circulate heat, the lunar surface heats up to about 127°C (above the boiling point for water) during the "day" and then plummets at "night" to −173°C. Living in a well-fortified shelter is one thing; venturing outside in just a spacesuit is another. NASA timed the Moon trips for the lunar "dawn," before the surface heated up to unbearable levels. These terms are in quotation marks because a complete "day" on the Moon—the time it takes to make a full rotation on its axis with respect to the Sun—is about 29 days, or one month on Earth. (That's not coincidental: because the Moon is in a tidal gravitational lock with Earth, its rotation on its axis matches its revolution around the Earth.) So, at the lunar equator, lunar dawn lingers for a few Earth days; high noon starts at the seven-day mark; sunset is at the fourteen-day mark; and nighttime darkness lasts for two weeks.

This means that, for most of the lunar surface, it is often either too hot or too cold to be outside, however protected. The sweet spot for outdoor work is lunar dawn and dusk, about a week out of every Earth month. A nighttime low of −173°C far exceeds anything experienced on Earth. The lowest natural recorded temperature on Earth was at the Soviet Vostok Station in Antarctica: −89°C (−128°F). As for the typical 127°C (260°F) high on the Moon, that's a slow-cook level suitable for stewing beef. This doesn't lend itself to efficient mining operations. Robots may be able to survive these temperatures to some degree, but not humans. The two weeks of darkness also means that no solar energy can be collected. There is no practical way to collect and store enough solar energy in the two weeks of light to last during the two weeks of darkness. Another energy source—perhaps a nuclear fission reactor, at first—will be needed. The prolonged darkness greatly limits the ability to grow vegetables, too.

The lunar day–night cycle, more than any other factor, is leading space explorers to establish bases at the lunar poles. Indeed, ESA's inflatable dome is planned for the Shackleton Crater near the lunar south pole, as is China's habitat. There, the temperature range and amount of daylight are nearly constant. Unlike Earth, the Moon doesn't have seasons. Its axial tilt to the ecliptic, or the Sun's apparent motion, is only 1.5 degrees, compared with the Earth's 23.5-degree tilt. Sunlit regions at the lunar poles average around −50°C, cold but workable, comparable with South Pole temperatures on Earth. More important, some high-altitude polar regions are sunlit 85 to 100 percent of the time. Malapert Mountain, an unofficial designation of part of the ridge of Malapert Crater, near the Shackleton Crater, may be a "peak of eternal light," a point on an astronomical body that always has sunlight. Nevertheless, even if locations fall into darkness for a day or two per month, just a few solar collectors thoughtfully positioned around the rims of a polar crater could generate a constant flow of electricity. This perpetually lit area will be prime real estate, indeed, despite the Space Treaty's premise that no one owns the Moon.

The added bonus of a lunar pole base is the ice in perpetually shadowed craters, just down the slope from where you are living. Of course, the engineering needed for tapping water that has been deeply frozen for billions of years in craters kilometers deep is far from trivial. And by "water" I mean frozen gravel that might only be 5 to 8 percent ice crystals. As a comparison, Sahara sand contains 2 to 5 percent water. This notion that the lunar poles have ice may engender visions of vast Arctic ice sheets ready for the cutting and harvesting. But this is not ice you can skate on. A wheelbarrow full would yield only a gallon or two of water—an oasis in relative terms only.[25] We could excavate the gravel and extract the water, or we could direct heat remotely to volatize the gravel and collect the vapor. We then would water use this for drinking, growing food, and also splitting into oxygen and hydrogen. Solar energy should provide enough power for this electrolysis. One promising note is that reanalysis of decade-old data from the Moon Mineralogy Mapper on India's Chandrayaan-1 spacecraft reveal that some shadowed polar creators might have ice at the 30 percent concentration-by-weight level a few millimeters deep.[26]

The process of electrolysis is straightforward, but only if it is feasible to excavate or volatize the frozen gravel, which hasn't seen sunlight in hundreds of millions of years and which is an estimated 40 degrees above absolute zero (−233.15°C, −387.67°F), perhaps colder than machines can handle. This might require nuclear-powered machinery, and special permits would be needed by any company wanting to launch these, let alone place them on the Moon.

Regardless of the profitability of lunar ice mining, the lunar poles are a practical place to set up bases for reasons of temperature stability and dependable solar energy. We could perform various science and engineering experiments here. That's a good start. But as fate would have it, much of the juicy stuff on the Moon, aside from water, that could prove to be profitable—helium-3 and KREEP—is in greater concentrations in the drier, less hospitable equatorial regions.[27] Minerals abound on the far side, too, a lonely place where the Earth will never, ever be visible.

There's the old joke about a drunk searching for his house keys at night under a streetlamp. When asked what he's doing, he explains that he lost his keys up the street. "So why are you looking for them over here, then," the joke goes. The drunk answers, "Because this is where the light is." Shift that to the Moon. Although more prospecting needs to be performed, the general consensus is that polar regions offer fewer mining opportunities, particularly helium-3, which is deposited from the Sun in its greatest abundances at the lunar equator. But we'll likely need to start mining operations in the less-rich polar regions because, quite literally, that's where the light (and water) is.

How We Will Live on the Moon

Science bases and camps on the Moon will be rustic at first. As a parallel to Antarctica, they will accommodate only four to eight people. We will need to import all life-sustaining resources: air, water, food, and warmth. As more nations and companies become interested in lunar real estate, these bases will grow, grouped by national identity in relative proximity. And like Antarctica, tourism will follow shortly after.

While it does offer some gravity, life on the Moon is nevertheless packed with the hazards of life in orbit. This dictates what form the shelter we must have. Science fiction writers often fantasize about massive, glass-domed cities. Yes, the warm yellow glow emanating from those domes sure seems homey, like a celestial nightlight. But such structures are impractical and unlikely to be realized for two reasons. First, as discussed previously, the Moon is bathed in deadly radiation from the Sun and beyond. For the most part, that which lets in light also lets in this radiation. On Earth, high-tech manufacturing using metals and sophisticated filtering materials might yield a glass that protects against harmful radiation while allowing optical light to penetrate. How do you make such glass on the Moon? Next, even if you do have the glass, and if it is also strong enough to withstand the impact of micrometeoroids that rain down regularly on the Moon, supporting such a structure will take a herculean engineering effort. The largest glass structures on Earth are no more than a few dozen meters across. Although low gravity allows you to build bigger on the Moon, practicality—that is, labor costs, building materials, workplace hazards such as −50°C cold coupled with no air or air pressure—trumps world-class designs. Decades of mastering construction on the Moon would need to transpire before architecture-as-art makes a debut.

This needs to be understood as a basic principle for all space settlements, with all due respect to space illustrators: first comes igloos and yurts, then more spacious quarters with modest charms, then the Taj Mahal only if advanced artistry can be supported by a building infrastructure, basic physics, and a purpose for such a grand design.

And concerning these massive domed cities on the Moon: whom are they for? Who will live on the Moon? The size of the lunar population will be dictated by gravity. If 0.16G is not enough force to allow for proper gestation of a fetus and subsequent infant and child development, then no one can raise a family on the Moon. Period. End of settlement. The Moon would be limited to an industrial park and science wonderland with some elements of tourism and maybe retirement. That, in turn, dictates the modest architecture for these transient lunar dwellers.

Space-Age Cavemen

The first Moon bases and camps, resembling Antarctic camps, will be located coincidentally at the lunar poles, where polar temperatures are nearly identical to those at the Earth's poles in winter. Harsh conditions—temperature and radiation—mean that workers will spend only a few hours per twenty-four-hour cycle outdoors and live in small domed habitats. Inside their huts, life would be much like it is now on Antarctica, only with less room—or like the ISS, but with more room. They would need to maintain a vigorous exercise regime to combat the effects of 16 percent gravity. And they will have busy days filled with infrastructure building or science experiments. One perk would be the fantastic view of Earth, appearing six times the size of a full moon. A hassle will be the toxic lunar dust, which sticks to everything because of electrostatic charge.

If history can serve as a guide, the lunar workforce likely will be predominantly male. Such is the case today in remote mining and science facilities. Few women ventured to Antarctica before 1970; by 1980 the ratio was about twenty to one, male to female, although by 2015 women made up about a quarter of the population, a rough estimate considering that people come and go so quickly. Gender equality is reaching some scientific fields these days but not so much in the mining and resource extraction industries that might come to define the Moon. Will the Moon become a lawless boomtown, as witnessed in the Bakken oil-producing region of North Dakota, where the influx of male workers between 2006 and 2012 brought crime, violence, heavy drinking, illicit sex, and other unsavory practices? That's mere speculation, but Antarctica sees its share of activities less pure than the ice that surrounds the bases.

I don't mean to paint potential lunar dwellers as a bunch of troglodytes, but as their numbers grow, they may very well be living in caves, like a few of our ancestors did more than 100,000 years ago. More accurately, these would be lava tubes, subsurface tunnels carved out billions of years ago by molten lava during the Moon's early development. The chances of finding a lava tube where we want to be—near water or other valuable resources—are slim, especially in the first decade after we return to the Moon. So, simple bases and human-made underground connections will need to serve us in the near future. But in terms of accommodating hundreds or even thousands of workers, lava tubes may be the way to go, particularly if the lunar transportation infrastructure is robust enough to move workers where they need to be.

As with caves on Earth, lava tubes can be conveniently spacious and flat, with some estimated to be hundreds of meters wide. These would be near-ready-made shelters from radiation, meteoroids, and, to some degree, temperature. The tubes may only get as cold as −20°C (−5°F).[28] All we would need to do is seal and pressurize them to enjoy a normal, albeit subterranean, life. We would need to rely on artificial lighting, but these lava tubes do have natural skylights. NASA's Lunar Reconnaissance Orbiter has imaged hundreds of holes that appear to be skylights into subsurface caverns. Japan's SELENE lunar orbiter has found one subsurface structure in the Marius Hills region of Oceanus Procellarum in the lunar northern hemisphere that appears to be a cave fifty kilometers long and a hundred meters wide.[29]

Some of these natural holes could be sealed to create a massive, pressurized shelter. The Mare Tranquillitatis Hole, a few hundred kilometers from the Apollo 11 site, is one possible location for a camp or even a lunar hotel. Approximately 90 meters across and 107 meter in depth to a flat surface, a bit like a sports arena, the hole appears deep enough that dwellers at the bottom would be protected from direct solar radiation and most cosmic radiation, while mirrors could reflect sunlight to illuminate the subsurface area during the two-week lunar day.[30] The bottom appears to have a positively livable day–night temperature range of −20°C to 30°C (−5°F to 85°F).[31]

We shouldn't forget the fact that we need to breathe. Although the Moon is airless, the lunar regolith is oxygen-rich, upward of 45 percent oxygen by mass, albeit locked away in minerals such as SiO_2, TiO_2, Al_2O_3, FeO, and MgO. The mining of these minerals would liberate oxygen to breathe. This would be important in the non-polar regions, where water (and the oxygen it contains) is less plentiful. NASA is testing in situ resource utilization (ISRU) methods to generate oxygen. One would be splitting H_2O into H and O, should water be plentiful. Another is utilizing the carbon dioxide / oxygen cycle with greenhouse plants, described further below. A third is heating the regolith to about 900°C and mixing it with (imported) hydrogen to generate water and then oxygen.[32] One such project is the Precursor In-situ Lunar Oxygen Testbed

(PILOT), which was built to scale in 2008 to achieve production rates equivalent to 1,000 kg O_2 per year to support a lunar outpost.[33]

Lunar Disneyland

Now to have a little fun. Tourism is certainly in the Moon's future.

One can easily envision a two-week trip to the Moon to be much like an African safari was 150 years ago: initially for the wealthy, with a tinge of danger, and certainly not for the kids, at least not at first. Several companies already are planning possibilities. One type of trip would be a flyby. This may happen soon, by the early 2020s, after tourism in low-earth orbit is established, because flying to the Moon but not landing is not much more difficult or expensive then launching to orbit a few hundred miles above Earth, to a space hotel. In 2018, SpaceX announced its first paying customer for the trip, Yusaku Maezawa, a billionaire businessman from Japan who made his fortune in online clothing sales. Maezawa plans to bring along at least five artists for a project called Dear Moon. The five-day trip, aboard SpaceX's BFR, now in development, may happen by 2023.[34]

Let's fast-forward a bit, though, to envision what real tourism might look like by the middle of the twenty-first century.

There will come a time when travel to the Moon is safe enough, comfortable enough, and affordable enough—all relative terms here—that the Aspen ski crowd, Davos attendees, and similar ilk will be able to visit the lunar surface and do touristy things. This has long been a goal of hoteliers. Barron Hilton, son of Conrad, pitched the concept of Lunar Hilton at a meeting of the American Astronautical Society in 1967, two years before Armstrong and Aldrin walked on the Moon. He had designs, a mock room key, and even a reservation card, clearly not envisioning the day when reservations would be made via the internet. (And what's a key?) Is discussing this hotel concept now as silly as discussing it then, a half-century ago? I'll boldly say no, because the technology is at long last coming together to make this a reality.

As stated earlier, constructing a large dome for a hotel will be a challenging engineering feat for the Moon. A more reasonable approach would to have a modest, subterranean hotel with several observation domes where you could visit for a few hours at a time at the expense of a small radiation exposure. Windows are a must, I agree. The view from the Moon is one of its three basic charms, for a tourist. The other two charms are the low gravity and the historic sites. First, the view.

"Magnificent desolation" and "more desolate than any place on Earth." That's what Buzz Aldrin has said about the Moon. This may not sound like a selling point, but there's certainly beauty in the desolate landscapes of the Badlands National Park in the United States or the Sahara Desert in northern Africa. The only difference is that the entire Moon is this desolate landscape, draped under a black sky. The landscape largely comprises lunar maria—so named because they look like oceans from Earth—which are large bone-dry plains formed by ancient volcanic outflows; lunar highlands, or terrae, which are featureless plains at a higher altitude than the maria; and impact craters, created by asteroid and comet collisions long ago. There's certainly a wow factor in standing on the edge of a massive lunar crater and thinking about the impact that created it.

But, the real beauty on the Moon may be the Earth, which would appear as a massive orb in the sky, up to six times larger than our view of the Sun or Moon from Earth. From a lunar perspective, the Earth will wax and wane through the course of a month, and swirling cloud patterns and large storms would be visible from your

Earthrise as seen from Apollo 8 as the crew orbited the Moon. Taken by crewmember Bill Anders on December 24, 1968, this now-iconic image was among the first photographs of the Earth from space, capturing the beauty and fragility of our home planet.

hotel observation deck. The view would be so breathtaking that it may be quite difficult to limit your time under the glass. And Earth will always be visible, day or night. This vista is unique, compared with any "colony" past or future—an ever-present direct view of the motherland, assuming you are on the near side of the Moon.

Because the Moon is airless, stars will not twinkle or be distorted. They will appear slightly bigger than they do from Earth, with more color. Also, because there's minimal light pollution, you would see many more stars with the naked eye—perhaps a view comparable to what most humans experienced in the era before electricity. There will be predictable solar eclipses, too, with the Earth passing in front of the view of the Sun, a spectacle than no human eyes have yet witnessed.

As for the low gravity, activities would be mind-blowing. In the safety of a pressurized hotel, you could literally fly by flapping artificial wings. This would take some strength, and practice, but it is certainly theoretically possible as far as the arms being able to generate enough flapping force to stay aloft in the 0.16G environment. You could jump ten feet in the air. This isn't dangerous, because you would fall more slowly back to the surface. You could also lift at least six times the weight you could on Earth. You could lift and throw your companion with one arm. Much of the fun on the Moon will be playing outrageous sports. Combine low gravity with a trampoline, and you could reach spectacular heights. One odd sensation, though, will be that everything tends to move more slowly. In low gravity, you roll down an incline at about a sixth the speed you would on Earth. Juggling would be easier.

Harrison Schmitt reckons one could ski reasonably well on the Moon. He moved about the Taurus-Littrow valley with the motions of a cross-country skier, opting for that technique instead of the famous lunar bunny hop. He even called out to his partner, Eugene Cernan, "Too bad I don't have my skis!" and proceeded to make skiing motions and noises. Schmitt envisions cross-country skis gliding efficiently across the terrain, as well as ski slopes down certain craters exposed to light, where it wouldn't be too frigid. The chances of crashing into a tree would be nil.

Now that we're playing outside, admittedly an unhealthy scenario given the dangers, note that you could outrun or outski the sunset. Because the Sun sets so slowly, if you move faster than 16 kilometers an hour at the equator in a westward direction, you'd stay a few steps ahead of the darkness. Sure, you'll never keep this pace for long, but on Earth you need to be on an airplane to outrun the Sun. Maybe you can impress your friends at the lunar ski chalet with that fact.

The initial tourists will likely stay at the polar camps. A clear disadvantage for tourism here would be limited views of the stars in the near-perpetual light and being far from historical sites, namely the six sites visited during the Apollo missions. How tempting it would be to build a hotel in the Mare Tranquillitatis Hole near the Apollo 11 landing site called Tranquility Base. There's much to see there: the first human footprint on the Moon; the unwavering US flag; and the descent stage of the Eagle lunar landing module with an attached plaque proclaiming, "Here men from the planet Earth first set foot upon the Moon. July 1969, A.D. We came in peace for all mankind." All told, about a hundred objects were left behind from the first Moon landing, mostly mementos, tools, and gear. The Lunar Legacy Project at New Mexico State University maintains an exhaustive list of every known item, along with a map—all on prime land whose ownership will challenge the Space Treaty.

The other five original landing sites will be of great historic interest to American tourists; the three moon buggies are still sitting where they were parked. And surely the unmanned (and sometimes intentionally crashed) landing sites for craft from Russia, Japan, Europe, China, and India will be sites to visit for citizens of those countries. The Apollo 12 landing site, Mare Cognitum, in the vast Oceanus Procellarum region, will be a place of particular intrigue. Legend has it that the so-called Moon Museum is there. This is an inch-size ceramic wafer onto which six artists—Robert Rauschenberg, David Novros, John Chamberlain, Claes Oldenburg, Forrest Myers, and Andy Warhol—each carved a simple drawing. Warhol's image is a penis or rocket ship. The wafer is said to have been tucked in a wrapper around one of the lunar lander's legs. The wafer is real; and it certainly got into the hands of a contracted NASA engineer, still unnamed. The question is whether the engineer truly placed the artwork in the lander. He told lead artist Myers that he did.

Getting Around Town

Maintaining hotels on the Moon will be no trivial endeavor, primarily because of the need for workers and supplies. The workers might live on the Moon, as they do in Antarctica, for a year or two at a time, depending on the seriousness of health effects from low gravity. They could lead the tours. Or, the small crew who brought you to the Moon might double as the tour guides. Moving about the Moon from one historic landing site to the next, or to and from lava tubes and other shelters and villages, will require advanced infrastructure. Airplanes can't fly on the airless Moon. Jetpacks aren't out of the question, but they use a lot of fuel. Rovers and buggies move at golf-cart pace, not ideal for traversing hundreds of miles of sandy, hilly terrain unless you have lots of time and food and protection. Maglev trains would be ideal. Encountering no air resistance, these could move as fast as airplanes. The train cars would be pressurized and quite comfortable. The rails could be forged from local iron. But considering the labor needed to forge iron and lay track in a hostile environment, the task will take years.

Lunar transportation will start simple, with pressurized rovers on wheels. The Apollo-era buggies—properly known as Lunar Roving Vehicles or LRVs—topped out at 16 km / hour (10 mph). Going much faster may be difficult because of the reduced traction in 0.16G and the lunar dust, which flies up into the wheel wells and undercarriage of the vehicle. You don't want to break down on the Moon hundreds of miles between bases. That would be tantamount to death as supplies run out and the heat of the day or cold of the night approaches. Slow and steady big-wheel rovers with enough supplies for two weeks are in order and would need to be delivered from Earth. Transporting these vehicles to the Moon will come at a great expense; and for the first decade or so, it's conceivable there will be only a few available and in high demand.

One workable and relatively simple transportation idea is a suspended cable system, harking back to the ski resort reference. Gondolas have the advantage of being far above the toxic dust level. They can move rather quickly, too, because there's no air resistance or crosswind. Unlike roads, which require concrete (and precious water to make that concrete), or rails, which require miles and miles of iron and ties, the cable system would need only supports stationed strategically across the lunar surface. A more far-out idea is the lunar hopper, difficult to describe because they don't yet exist. The idea is for a vehicle with four springy legs that snap to propel the coach across a great distance, much like the leap of a flea. The low gravity means the coach lifts easily and lands softly.

Until transportation infrastructure is in place, lunar workers will be confined to a radius of just a few dozen kilometers around their base, what could be reached in a few hours with a moon buggy, jetpack, or simple device. This puts a serious crimp, and delay, into tourism development, let alone grand designs of living there comfortably for any length of time. In short, you'd be confined to your base, much like someone overwintering in Antarctica.

Unsettling Health Concerns

As noted, living on the Moon exposes the lunar dweller to many serious health risks. Many can be mitigated with proper shelter. The low gravity, however, will very likely be the showstopper for a civilization to develop on the Moon. We covered gravity in Chapter 2. In low gravity for prolonged periods, visitors to the Moon run the risk of never being able to return to what will feel like the crushing gravity of Earth. There's perhaps a point of no return, not known to science. Maybe after ten years on the Moon, your body may never readapt to life on Earth. Your bones might shatter upon return. Nothing, absolutely nothing, is known about the long-term effects of partial Earth gravity.

I discussed the concept of orbiting cities in Chapter 3. Such dwellings in space offer many more advantages than a lunar-based existence because there, you can manipulate gravity. You can rotate the entire orbiting contraption to mimic the feeling of the force of Earth's gravity. Once you are on the surface of the Moon, though, creating a rotating world will be too complex to be practical. You would need to create a Tilt-A-Whirl the size of an indoor mall, angled in such a way that riders—that is, lunar citizens—are feeling the centrifugal force at their feet, not pinning them against the wall. Such schemes are out there; the animation makes it look feasible.

Again, it's all about the "why." Why live on the Moon with no air or water or vibrant colors just to be in an underground Tilt-A-Whirl with limited ability to explore the terrain—for a lifetime, with kids to raise? What is the attraction? A visit, yes. But the novelty will quickly wear off. How will hundreds of thousands or millions of people be attracted to this life to populate the Moon under these conditions? Extreme poverty on Earth is a miserable existence, but at least you have gravity and air.

Aside from the low gravity issue, we will have great difficulty protecting ourselves against the toxic lunar dust. Actually, dust is a poor word for this substance, implying something fluffy and easy to wipe off. Lunar dust is razor-sharp and abrasive like asbestos. The dust got this way because of a constant bombardment of micrometeoroids and radiation that has chiseled the regolith into micro-weapons, like early humans pounding rocks into spear tips. And the lack of liquid water on the Moon assures that the regolith would never be eroded into smooth curves.

Scientists have found that the dust will slice through and kill lung cells or cause irreparable DNA damage if inhaled, a sort of *lunar regolithosis,* if I can coin a term.[35] Apollo scientists suffered from dust exposure even though they spent such a limited time on the lunar surface. Harrison Schmitt accidentally breathed some in and experienced the symptoms of hay fever for a day. The dust is difficult to avoid because it is statically charged, clinging to everything and yet slow to settle when disturbed in the low gravity. Even more ominous, there appears to be dust fountains or streams a few feet of the ground as a result of a phenomenon called electrostatic levitation. So, even if you don't play in the dust, you will collect dust on your spacesuit when merely walking outside. You drag it inside the habitat with you.

Eugene Cernan, the last man to have walked on the Moon, captured the enormity of the dust problem in a technical debrief of the Apollo 17 in 1973: "I think dust is probably one of our greatest inhibitors to a nominal operation on the Moon. I think we can overcome other physiological or physical or mechanical problems except dust."[36] The dust even jammed the astronauts' zippers, jeopardizing the integrity of their spacesuits. Spacesuits and equipment might have failed if he and Schmitt had stayed on the moon for a little longer. Since this was not documented until the longest but final Apollo mission, NASA doesn't yet understand the level of protection needed to keep astronauts safe, let alone any long-term lunar inhabitants.[37] This is in contrast to solar and cosmic radiation, against which we know how to protect ourselves. Protection from lunar dust likely will require a decontamination and exhaust system similar to those used for asbestos removal. That's life on the Moon for everyone.

Growing Food on the Moon

Next up on the hazard menu, food security. As with most other activities on the Moon, food production will need to be sheltered—that is, growing food not in the ground but rather underground. Scientists and engineers have made remarkable progress on this challenge, though. For example, the Controlled Environment Agriculture Center at the University of Arizona has created a tube-shaped lunar greenhouse about 5.5 meters long and 2.2 meters high that can produce about 1,000 kcal of food daily. This greenhouse, a bioregenerative life support system (BLSS), could additionally potentially provide 100 percent of an astronaut's air and potable water. In this closed loop system, plants are grown hydroponically under lights, nourished by nutrient salts containing the necessary nitrogen, potassium, and phosphorus for plant growth. With that input, the plants transpire oxygen and water vapor for the astronauts to breathe and drink; the plants themselves, of course, get eaten. The astronauts, in turn, exhale carbon dioxide, urinate, and defecate, and these products are fed back into the system. In short, by creating a miniature biosphere and placing this within the lunar habitat, you wouldn't need a bulky O_2/CO_2 exchanger, as you do on a submarine or on the ISS. Your food is your air and potable water.

According to Gene Giacomelli, who led the development of the BLSS, the system is capable of producing 26 grams of fresh edible biomass per kilowatt-hour, so efficient that it could run on electricity generated by solar panels, at least during the lunar day. Plants exposed to natural surface light would be inundated by harmful radiation, just like humans. One could design underground

The Prototype Lunar Greenhouse. Designed by University of Arizona's Controlled Environment Agriculture Center, this 6- by 2.5-meter cylindrical chamber is lightweight and expandable and could be used underground on the Moon or Mars. The prototype is designed to provide 100 percent of oxygen needs for one person, as well as 1,000 kcal of food daily.

greenhouses that are fed reflected solar light at the lunar poles, where the Sun shines at least 80 percent of the time. These would require serious construction, though, in the form of a massive excavation of regolith to create the kind of greenhouse seen on Earth outdoors. Maybe once a lunar infrastructure is created, large greenhouses like this could be built. The beauty of the BLSS is that they are highly efficient in maximizing the space-to-crop ratio by stacking crops and dangling light where it is needed. Also, they are inflatable and relatively light to ship to the Moon.

Giacomelli's group has designed a four-person lunar habitat that incorporates four BLSS, with the dual goal of plants providing food and oxygen. More food would need to be flown in. A hard-working astronaut would likely exert 3,000 kcal per day. A change of grow lights would be needed from time to time, too, along with nutrients. Still, reducing the food shipment by one-third would save the early lunar missions money and provide the invaluable psychological lift of having fresh food on hand, to be supplemented by grains and other "pantry" items. The BLSS prototypes on Earth grow leafy greens, strawberries, tomatoes, and sweet potatoes entirely indoors. They can be more efficient than plots grown outdoors, because the University of Arizona team can control the temperature, humidity, nutrient feed, and pests. No weeding!

The University of Arizona BLSS is just one prototype, but NASA ultimately will use this or a similar, self-contained system if they choose to go in the direction of supplemental, localized food production, as opposed to 100 percent food delivery. China's Lunar Palace 1, mentioned earlier in this chapter, includes a variety of important crops for protein and fat, such as soybeans. In Chinese news reports, the research team claimed that the system was 98 percent efficient, meaning only 2 percent of supplies were added into the system once the experiment started, which would be impressive if verified.* Other teams worldwide are attempting to grow food in lunar regolith simulants.[38] This hasn't worked so well; plants germinate but fail to thrive. One must wonder the purpose of perfecting growth in regolith, though, if hydroponics is superior for efficiencies in space and water allocation. A

*Few technical details of the project are available, and no member of the science team replied to my query. No discernible airlock is visible, so the integrity of the O_2/CO_2 cycle is unclear.

variety of crops, even root vegetables, can be grown in a hydroponic system. And regolith simulant is not regolith; it would make more sense to prepare for the Moon with a working hydroponic system and then, once on the Moon, experiment in other forms of farming.

Moreover, the Moon's proximity to Earth, coupled with such a high hurdle for perfecting large-scale lunar agriculture to feed the masses, likely will result in never *having* to be 100 percent self-sufficient in food production. It may be easier to fly most of the food in, as we do now for Antarctica. Having a large population on the Moon would imply that access to the Moon has become cheaper, so imported foods might not be that expensive. To ramp up food production to feed thousands of people, we would need thousands of BLSS, each maintained by its owner. Or, more traditional greenhouses would need to be set underground with intricate mirroring for reflected natural light to reach all the plants. The lack of carbon, nitrogen, and hydrogen on the Moon would imply that these atoms would need to be imported, at least once, in massive quantities, to provide the basis of the materials needed for agriculture. Yet another hassle is the fact that pollination probably would need to be done by hand, because bees and other insects might not be able to fly in low gravity in the absence of a magnetic field, let alone reproduce and thrive. The 0.16G would likely rule out animal husbandry. Growing fungi is possible if enough water is found. Same goes for algae grown in vats. Algae, yum.

Terraforming the Moon

Living on other moons and planets will always have the feel of being trapped in a hotel or, at best, an indoor mall. You will never, ever experience the sensation of being outside. There is no outside even when you're outside, because you are confined to a spacesuit needed for air, pressure, or radiation protection. No wind in your hair, ever. Maybe you can have a fan blowing indoor air in your hair, but never real wind—or real rain—unless we terraform the place.

Terraforming the Moon—that is, making it a mini-Earth so that you can walk outside without protective gear—is a plan for centuries to come, at best. The entire prospect of terraforming the Moon is a marriage of brilliant engineering and poor logic. At a minimum, we would have to create an atmosphere. The primary method would be to deliberately bombard the Moon with fifty to a hundred comets, taken (somehow) from the edge of the Solar System. This could simultaneously add oceans' worth of water and a lot of nitrogen while liberating oxygen from the soil into the air and increasing the rotational speed to a twenty-four-hour day.[39] But such orchestrated destruction requires evacuating the denizens of the Moon for a century or so as you renovate it and then repeating this process every thousand years or so, because the Moon doesn't have a magnetic field and thus no protection from solar winds, which would blow away most of the atmosphere in relatively short order. And if we had the technology to do all this, we'd likely instead be building massive orbital cities where we could control the level of gravity. To be clear, terraforming the Moon still won't fix the gravity problem. Terraform Mars, sure. That's neat, as well as possible, over the course of centuries. Terraform the Moon? That's silly and off the table for discussion until a technology we cannot envision today allows us to quickly terraform the planet without disrupting the lives of those already living there.

An industrialized Moon with transient workers, dependent on the Earth for food and other supplies, is the direction I see. But good things take time. We can't simply jump over to the Moon and start up just because we've been there before. Before humans return, it would be wise to have basic infrastructure in place, such as lunar communication satellites to enable astronauts to talk to each other at distances farther than their local horizon. Telecommunication companies are vying to establish a 4G cellular network for the Moon, provided there's a market to justify the expense.[40] A series of orbiting lunar habitats could provide a valuable safety net, too, for supplies and evacuation. Our bases on Antarctica are robust now because such infrastructure is in place, enabling the bases to serve a purpose without being extraordinarily expensive to maintain. It took about fifty years after the first explorers set foot on that ice continent before bases were established, and another fifty years before access became routine. In this regard, fifty years after Apollo, we are right on track to return to the Moon for good.

My prediction: Boots on the Moon by the late-2020s; rudimentary robotic prospecting by the end of the 2020s; lunar orbital tours for wealthy clientele by the end of the 2020s; permanent human presence in the form of small science and mining camps by the 2030s; minor tourism and hundreds of people living temporarily on the lunar surface by the 2040s; access to the Moon for vacation or study affordable to most people by the end of the twenty-first century.

A final note on the Moon: I can't imagine that anyone reading this book would subscribe to the Moon hoax theory, the preposterous notion that NASA faked the Moon landing and made all those videos in a secret warehouse. Believers in the hoax are surprisingly plentiful. And you simply can't argue scientific facts with them. But here's a bit of logic they can't brush aside: the race to the Moon was a competition with the Soviets; and if the United States had faked it, the Soviets would have cried foul. They didn't. They congratulated the Americans.[41]

5

Living on Asteroids

Asteroids. Some call them space rocks; others, veritable gold mines. They are mostly the shattered remains of planetesimals, solid bits of debris comprising minerals and metals that never formed into planets during the early days of the Solar System. The vast majority orbit the Sun in a beltway between Mars and Jupiter. Some asteroids, called Jupiter Trojans, circle the Sun in Jupiter's orbit; others, the near-earth asteroids (NEAs), swoop in rather close to us, occasionally coming closer than the Moon. One NEA, 350-meter-wide Apophis, may approach to within 31,000 km from Earth in April 2029; that's below the orbits of geosynchronous communication satellites. Yikes!

Asteroids differ from meteoroids, which are rocks less than a meter across. Asteroids are rocks larger than a meter. There are hundreds of millions of them larger than a hundred meters and likely billions all told. The largest, named Ceres, is so large that it is now classified as a dwarf planet like Pluto, approximately 1,000 kilometers across and making up nearly a third of the mass of the entire asteroid belt. The next eleven largest asteroids make up an additional third of the total asteroid mass, with the remaining untold billions—crumbs, really—constituting the remaining third. A common trope seen in sci-fi movies is the perilous flight through the asteroid belt, dodging boulders left and right. Truth is, space is large, and these asteroids are mostly separated by hundreds of thousands of kilometers of three-dimensional space. You'd have to be a rather lousy pilot to crash into one. However, the eighth-largest asteroid in the belt, Sylvia, has two tiny "moons" orbiting it—Remus and Romulus—which are about 700 km and 1,300 km away, respectively. Only one object in the asteroid belt is generally visible to the naked eye, and that's Vesta, about half the size of Ceres but closer and more reflective of sunlight.

Asteroids also differ from comets, which are "dirty ice balls" formed much farther from the Sun, at the edge of the Solar System.[1] Some comets can have highly eccentric elliptical orbits that bring them relatively close to Earth in cycles of about one hundred years. As they get close to the Sun, the ice and volatile gases burn off, creating the characteristic tail of the comet visible to the naked eye. I'll discuss comets more in Chapter 7; they will be useful in centuries to come if we colonize the outer Solar System.

Asteroids might not have the lure of the Moon or Mars, but they are valuable resources nonetheless. Many contain trillions of dollars' worth of precious metals, such as gold and platinum. And because they have only weak gravitational fields, they require little energy to land on and leave, almost like docking with the ISS. Should we expand our reach into the local Solar System, asteroid mining can play major role. They are worth their weight in gold for the just the water they contain, let alone the gold. In this regard, many scientists see asteroids as more viable for mining than the Moon. Indeed, there is an ongoing, heated argument between those who are pro-Moon and those who are pro-asteroid, as if the choice is one or the other. In my opinion, both types of mining options have challenges and rewards, and we really can't say which is better right now. (In other words, don't let anyone sell you on a business venture just yet.)

There are three main types of asteroids: C-type, S-type, and M-type. All of them are valuable. The C-type are carbonaceous, primarily carbon. They constitute the majority of asteroids, some 75 percent, and they are on the farther edge of the asteroid belt, closer to Jupiter than to Mars. At that distance from the Sun, they are colder and haven't boiled away their water into space. That water is locked up in ice; a good 10 percent of the mass might be water ice. Many C-type asteroids also contain phosphorus, which science fiction author Isaac Asimov called "life's bottleneck," because the amount of phosphorus on any given planet will determine how much life the planet can support. Mars has little phosphorus. So, humans on Mars may need to import it—from C-type asteroids. We could use that phosphorus on Earth, too. Some of these asteroids also have ammonia, which could provide Mars with much-needed nitrogen. Mars does appear to have enough water in the form of underground ice to support human life. But icy C-type asteroids towed into Mars's thin atmosphere, just skimming the top in a controlled maneuver, could off-gas lots of water vapor and oxygen, which could help build a livable Martian atmosphere. Ceres and Vesta are C-types, albeit far too large (and dangerous) to move.

S-type asteroids are silicaceous, or stony, rich in silicates such as quartz and granite. Silicates are needed for cement, ceramics, glass, and that essential stuff we call soil. S-types also contain nickel, iron, and precious metals. These make up almost 20 percent of asteroids. These metals could be mined for space-based constructions, such as solar panels and spacecraft. M-types are metal-rich and make up about 5 percent of asteroids. The "M" could very well stand for money. Some of these asteroids contain trillions of dollars' worth of platinum, gold, titanium, and other highly desired metals. Asteroid 16 Psyche, among the largest of the M-type—about 200 kilometers in rugged diameter—is thought to contain $10 quadrillion worth of iron, nickel, and gold.[2] That price estimate is a bit silly, because prices are set by scarcity, and hauling 16 Psyche to Earth would crash the market. Nevertheless, there are riches to be had if, or when, the price of retrieval drops below the price of metal on Earth. Private companies are eyeing M-types for platinum-group metals: iridium, osmium, palladium, platinum, ruthenium, and rhodium. These have many useful catalytic properties for Earth-based industries. They are rare in the Earth's crust, and what little is there came, in fact, from asteroid collisions.

Note that these basic classifications—C, S, and M—originated more than a century ago. Modern observations have broadened the classification system and have found overlaps in the early system.*

The gist here is that everything we need to live in space is contained in asteroids. Actually, we are already mining asteroids in that many of the metals we mine on Earth came from asteroids. Much of Earth's native stores of these metals is below the mineable crust. During Earth's molten youth, gravity pulled the siderophile, "iron-loving," elements down toward its core. These include cobalt, gold, nickel, silver, tungsten, and many more. Heavier stuff floated down; lighter stuff such as hydrogen, carbon, nitrogen, and oxygen bubbled up. What we mine near the surface is mostly what crashed down on Earth over eons. An added bonus in mining asteroids in space is that the materials have been segregated for us.

*Asteroids also can be categorized by orbit type: Amors (never reach Earth's orbit), Apollos (cross into Earth's orbit with period greater than a year), and Atens (cross into Earth's orbit with period shorter than a year).

Some asteroids are nothing but pure metal cores, with outer debris lost through the ages. Most never experienced a molten state, so precious metals didn't settle to the center of them. Some asteroids appear to have gold at surface concentrations as high as 0.7 ppm, compared with 0.001 ppm in the Earth's crust (which, again, is mostly gold dust left from a gold-heavy asteroid collision). Platinum may be as high as 63.8 ppm, compared with 0.005 ppm on Earth.[3]

Water Is the New Gold

If some of these resources sound mundane—water, ammonia, iron, and the like—you need to remember: what something is worth on Earth is different from what it is worth in space. How much does water cost? It's about ten cents a gallon on Earth but $10,000 a gallon on the ISS. Some asteroid resources will be precious and profitable in space, like water, and other asteroid resources, such as those in short supply on Earth, will be valuable down here. There also are exotic materials associated with asteroids, such a lonsdaleite, which is harder than diamond and can form when a chunk of asteroid carbon crashes onto Earth or, more safely, is directed to the Moon.

Several companies are developing technologies to mine asteroids as we learn more about where to "dig." The "where"—or, in this case, which asteroids—is a crucial question, given the expense of traveling to these asteroids with mining equipment. That's where governments come in, and we have an Earth-based analog here to learn from. Throughout human history, governments have funded expeditions to access the resources within their own borders. The Lewis and Clark expedition from 1804 to 1806 across the American west to the Pacific Ocean, commissioned by President

Thomas Jefferson, cost about $50,000 (several million dollars today). The expedition not only mapped the territory newly acquired from the French but also amassed information about timber, mineral deposits, and other resources needed for a growing country. In this way, modern governmental agencies are exploring asteroids and performing deep-space prospecting with telescopes and space probes, ascertaining their content via the spectral reading of their reflected light and off-gassing. Knowledge of the potential value of asteroids would, in theory, promote commercial mining.

Among the major "expeditions" to asteroids, all of which are robotic, is NASA's OSIRIS-REx (Origins, Spectral Interpretation, Resource Identification, Security, Regolith Explorer) mission, launched in 2016. OSIRIS-REx arrived at asteroid 101955 Bennu in December 2018 and will return a sample for analysis in September 2023. These things take time. It took two years for the craft to reach Bennu, a C-type asteroid about a half-kilometer wide, as wide as a skyscraper is tall.[4] Another two years will be spent orbiting the asteroid at an altitude of about five kilometers to determine where to land and sample; then the spacecraft will land for just a few seconds to scoop up that sample, up to 2 kilograms, before leaving for its two-year trip back to Earth.[5]

Bennu, discovered in 1999, is a near-earth asteroid that comes somewhat close to Earth (between Earth and Mars) every six years. It is considered a potentially hazardous asteroid, or PHA, because it may someday collide with Earth. NASA chose it for exploration primarily for scientific reasons. Bennu is relatively close, large enough to work with, and contains pristine carbonaceous material from the origin of the Solar System. The mission also is an engineering exercise in mining: how to orbit, land on, scoop from, and leave an asteroid. If the path is too close for comfort for these near-earth asteroids, we may someday soon be able to nudge it out of our way—or move it to a lunar orbit to mine at our leisure.[6]

NASA's OSIRIS-REx isn't the first landing on an asteroid. That honor goes to NASA's NEAR spacecraft, which landed on the asteroid Eros in 2001. The Japan Aerospace Exploration Agency (JAXA) was the first to collect a sample of an asteroid with its Hayabusa probe, which landed on asteroid 25143 Itokawa for about thirty seconds in 2005. Sampling didn't go as smoothly as planned, and JAXA wasn't sure what it had retrieved until a capsule returned to Earth in 2010. The samples weren't the chunks of rock that JAXA had hoped for, more like dust; but nonetheless, this constitutes the first returned samples from an asteroid. JAXA also demonstrated the use of ion-thruster engines, ideal for precise maneuvering on and around an asteroid. Hayabusa2 arrived at

The asteroid 162173 Ryugu. The Japanese probe Hayabusu2 reached Ryugu in 2018 with the goals of landing rovers, exploring the surface, and returning samples to Earth. Ryugu, rich in iron, water, and other resources, is approximately a kilometer wide. Seen here on the right, Hayabusu2 casts a shadow on the asteroid's surface as it approaches.

asteroid 162173 Ryugu in June 2018 for a similar exercise, this time with four surface mini-rovers, with plans to return samples in 2020. The rovers are like none you've ever seen before—no wheels. The gravity on Ryugu is so weak that wheels wouldn't be able to grip the surface. So, these rovers look like boxes that flip and tumble across the surface of the asteroid. The mission could place Japan at the forefront of asteroid mining.

It may seem contradictory that landing on asteroids, which I described as being as simple as docking with the ISS, has proven to be more difficult than landing and collecting samples on the Moon. The reason for the difficulty is threefold: asteroids are farther than the Moon, and communication to and from the craft can be delayed for minutes; the surface of the asteroid can be jagged, and it takes a lot of fuel to zoom around to find the best landing place; and these are missions running at bare-bone costs, far from the "money's no object" freedom of the Apollo era.[7] And practice makes perfect.

NASA's spacecraft *Dawn,* retired in November 2018, used ion propulsion technology to enter and leave the orbits of two celestial bodies, Vesta and Ceres, the first craft to do so. This is a notable example of how NASA introduces new technologies that serve its primary purpose—that is, space science—but prepare us for advanced space exploration. We'll need ion propulsion technology for asteroid mining and also for traveling to Mars.

As we get good at landing on asteroids, private companies are developing business plans to mine these rocks in situ, tow them closer to Earth, or provide services to enable other companies or governments to work in space. Planetary Resources, Inc. launched two satellites, in 2015 and 2018, respectively, to test the technology of low-cost telescopes for astronomy, Earth observations, and the finding and tracking of lucrative asteroids. In October 2018, Planetary Resources was acquired by ConsenSys, a company that develops blockchain software and tools for cryptocurrency. The company has speculated that cryptocurrency might be used for investment in space mining and subsequent selling of mined materials in lieu of an Earth-government-based financial authority.[8] Similarly, Deep Space Industries (DSI) is developing a business plan to make money in space, ultimately with a string of orbiting refueling stations selling water, oxygen, or hydrogen mined from asteroids. In the near term, however, DSI is building propulsion systems to enable a spacecraft to jump from low- to high-Earth orbits, or change its delta-v, at lower costs. This would open up more asteroids to mining. For example, at the orbit attainable with a delta-v of 4.5 km/sec, about 2.5 percent of NEAs are accessible. Ramp that up to 5.7 km/sec, and you can reach about 25 percent of NEAs.[9] In this way, DSI is helping build a commercial deep-space market.

Martin Elvis, a quasar expert at Harvard-Smithsonian Center for Astrophysics, has taken a strong interest in asteroid mining in recent years. He says that a small asteroid, about thirty meters in diameter, may contain 300 metric tons of water. That's enough hydrogen to fuel a mission to Mars from low-earth orbit. A private company may be able to fuel a NASA or ESA rocket for a cool $1 billion.[10] Many experts speculate that mining on the Moon—as challenging as that is—would be easier at first. Nevertheless, asteroids can provide inexhaustible resources once the technology matures. Private companies are eyeing them while simultaneously working to reduce the cost of access to space, the primary limitation for commencing with mining today.

Who Wants to Be a Trillionaire?

Who owns the asteroids? According to the Outer Space Treaty, no one and everyone. As noted in the previous chapter, the treaty is ambiguous and likely will be redrawn or at least reinterpreted to allow the commercialization of asteroid mining. Some people can argue convincingly enough that it is not fair for any company to profit from the Moon—to burn up all that precious water just for rocket fuel, for example—but that argument is harder to maintain for asteroids. There are just so many of them—literally one for each person on Earth—and they are lifeless, like the Moon. And given the limited resources on this abode we call Earth, it would be foolish not to tap into the resources space offers. Why would it be acceptable to mine an asteroid only after it falls to Earth, as we do now—de facto, billions of years *post facto*—and not mine them while they are in orbit?

Astrophysicist Neil deGrasse Tyson has said the first trillionaires will be those people who exploit the resources found in asteroids. This may be hyperbole. Billionaires, yes, definitely. Trillionaires? Well, as just mentioned, there are too many asteroids for any single person to monopolize. And simply seizing a trillion-dollar platinum-rich asteroid doesn't mean you will sell a trillion dollars' worth of platinum. Reclaiming that much platinum would cause prices to fall, making the mining less profitable. Should any space baron decide to hoard resources, someone else would simply find another asteroid among the billions. But there is plenty of truth in the fact that the masters of asteroids can establish a market for fuel and other resources as part of a vibrant space economy worth trillions.

The first asteroids to be mined will be the closest, essentially low-hanging fruit. Interestingly, these asteroids are considered both dangerous and valuable: dangerous because they are large enough and close enough to wipe out life should they collide with Earth; valuable because they are large enough and close enough to land on and mine.

The asteroid Ryugu, the one Japan is exploring, is among the most viable. Ryugu is about a kilometer across and contains abundant deposits of water, ammonia, cobalt, iron, and nickel. A kilometer might seem minuscule in the cosmic scheme of things, but try to form a mental image of an open-pit mine of that size on Earth, with a few trucks and diggers. Millions of tons of material can be extracted from such a small volume. Among Ryugu's materials, the water would be the most valuable at first. Ammonia would be important in a few decades for settlements on the Moon or Mars, which lack the nitrogen needed for large-scale farming. Practically speaking, nickel and iron are not in short supply on Earth and won't be needed in space until we start building big things up there, such as spacecraft and orbital cities. Asteroidal cobalt may soon be needed here, though, because there are *ethical* shortages on Earth, as cobalt mining so often entails exploitive labor practices in central Africa.

Other target asteroids are 1989 ML, Nereus, Didymos, 2011 UW158, and Anteros. The 2011 UW158 asteroid, discovered in 2011, is only 300 meters across and contains untold riches of platinum. I mention this only to underscore the fact that potentially exploitable asteroids are being discovered every year. Mining of these asteroids would be mostly robotic. Depending on their consistency, asteroids can be surface-scraped with an auger conveyor or raked with magnets, which would be done on asteroids that are mostly low-gravity, loosely held rubble. Others with volatiles such as ammonia or water could be heated to collect the vapor. Harder asteroids could be shaft-mined. Some small asteroids, a few tens of meters wide, could be towed to low-earth orbit and "parked" next to a space station for gradual deconstruction by workers living in space.

NASA's Asteroid Redirect Mission, or ARM, was to rendezvous with a near-earth asteroid, grab it with robotic arms or otherwise harpoon it, and tow it to cislunar orbit. The mission was canceled in 2018 under White House Space Policy Directive 1, with the new priority to instead go to the Moon. ARM would have targeted a very small asteroid, only a few meters across yet still containing several tons of material, and hauled it to a stable lunar orbit for astronauts to essentially mine—that is, chisel off samples and return them to Earth. One key technology tested would have been solar electric propulsion, which uses electricity from solar arrays to create electromagnetic fields to accelerate and expel charged ions, creating precise thrust with a very efficient use of propellant. Learning to manage large mass with low thrust will benefit future Mars missions. On that note, Mars missions could rendezvous with a small asteroid, crush it into a fine gravel, and coat the spacecraft with it to create a radiation shield, thus solving the problem of traveling to Mars in a well-protected vessel that would otherwise be too heavy to launch from Earth. The ARM was to also demonstrate planetary-defense technology, because the ability to tow asteroids toward Earth implies the ability to deflect asteroids away from Earth. As such, the mission sits in a NASA warehouse pending renewed funding.

Many technologies need to be created, let alone perfected, before asteroid mining is possible. These include precision maneuvering, anchoring onto a low-gravity object, and processing ore in near-zero gravity. There are trade-offs when considering mining the Moon or asteroids. The Moon is consistently closer, and you can establish large-scale and long-term operations. Gravity allows heavier and lighter elements to separate. The downside is that there is still a gravity well to contend with. Landing heavy equipment is difficult because of the 0.16G and lack of atmosphere for braking. Fuel is needed to fire engines to slow the descent. Launching, or exporting, would require a lot of fuel, too, until mass drivers are installed to fling material off the lunar surface. The lunar escape velocity is 2.38 km/s, about twice as fast as a speeding bullet. Asteroids are farther away, but some require less fuel to reach because of the low delta-v, the change in velocity needed to land on them. That's less fuel to burn to slow down. The asteroids that are only a few hundreds of meters wide have negligible gravity, and landing is akin to docking at a space station, as noted. Escape velocity is low, too, a few meters per second. You could jump off most asteroids; and, indeed, the mining equipment would need to be anchored so that digging (action) doesn't result in pushing the machines off the surface (reaction). With all of these challenges, mining in near-zero-gravity has never been accomplished. As we saw with the JAXA mission Hayabusa, even a quick scoop is challenging.

Enter Luxembourg

Yes, Luxembourg. It's among the smallest counties in the world. Small but rich, and they want to be richer. Luxembourg has invested hundreds of millions of dollars in space mining, including $28 million in Planetary Resources for a 10 percent share of the company.[11] Although Luxembourg signed the Outer Space Treaty in 1967, the country passed a law almost fifty years later to the day, in 2017, providing companies the rights to space resources they extract from the Moon, asteroids, or other celestial bodies.[12] The tiny nation has long been a leader in the satellite communications industry. Now it seems to want to be the Silicon Valley of space mining. For starters, Planetary Resources (now under ConsenSys) and Deep Space Industries have set up offices in Luxembourg. And China has signed an MOU with the tiny country concerning exploration and use of resources.

Luxembourg hopes to do for space mining what it did for communication satellites. Before the 1980s, communication satellites were government-funded or government-regulated, then the rules loosened. In 1985, the Luxembourg government supported the creation of a company called the Société Européenne des Satellites, the first private satellite operator in Europe and now the second-largest in the world, with more than sixty satellites in orbit. The country may have the financial backing, legal framework, and loose regulatory structure to make space mining a reality. According to Luxembourg deputy prime minister Etienne Schneider, who signed the law, "Our goal is to put into place an overall framework for the exploration and commercial use of resources from 'celestial bodies' such as asteroids, or from the Moon."[13]

The Luxembourg law is not without precedent. The United States passed a similar law in 2015, the Spurring Private Aerospace Competitiveness and Entrepreneurship (SPACE) Act, which states that private companies can own and sell what they extract; they just can't own the celestial body itself, abiding to the wording of the Outer Space Treaty. That's a bit of a paradox, though. Under this US law, a company could mine an entire asteroid, selling the resources as water, fuel, oxygen, or construction materials, until there was not a crumb left and thus never have owned what it completely cannibalized. Lawyers began debating the legality of this US law before the ink had dried. Some argue that the law is a direct violation of the Outer Space Treaty, Article II, which states "Outer space, including the moon and other celestial bodies, is not subject to national appropriation by claim of sovereignty, by means of use or occupation, or by any other means." At issue among lawyers are the terms "national appropriation" (should a private company be treated as a nation?) and "by means of use or occupation" (well, just being on the Moon means you are using it).

The Outer Space Treaty succeeded in keeping weapons off the Moon. No one challenged the commercial aspect of lunar and asteroid mining, though, because in 1967 this was fantasy. Now it's a reality, an entirely different era, which is why two countries so far have created laws to incentivize investment. I don't see how companies will invest in space without reassurance that profit from space resources is at least feasible. And I don't see how the Outer Space Treaty and, certainly, the Moon Treaty can allow for profits. So, I predict increased pressure to revisit the Outer Space Treaty as access to space gets easier.

Living on and in Asteroids

This is a book about space settlements, and I've been rambling on about robotic mining. There seems to be very little reason to live on an asteroid to mine it. Initially, human presence may be confined to brief visits to the asteroid belt to secure equipment or aid in transport of mined materials. The exception might be Ceres, which, as noted, is now deemed a minor planet. It's about half the size of Pluto and a quarter the size of the Moon. One could envision Ceres as being a scientific outpost or a supply hub to support a network of asteroid mines that service both Mars and Earth, occupied by temporary crews by the end of the twenty-first century. Then again, an orbiting spaceport in that vicinity might be more habitable.

Life on Ceres would be rough. Airless like the Moon, Ceres would have no natural protection from radiation. Workers would need to burrow underground for shelter. That's likely possible, but the composition and depth of the crust, and where it meets Ceres's ice layer, is largely unknown. The flight to and from there would take years, as it is beyond the orbit of Mars. Its gravity is 0.03G, not as crippling as microgravity but unlikely to be suitable for long-term human health. The sunlight there is ten times weaker than it is on Earth, so collecting solar energy is difficult. You would likely need nuclear energy to sustain any industrial operations. Communication with Earth would be delayed by an average of thirty minutes. To wit, living on Ceres would be more challenging than living on the Moon given the remoteness, lack of gravity, and scarcity of resources other than water and a few minerals.

There's good science to be had, though. There is a remote chance that Ceres could harbor alien life beneath its icy crust in a liquid ocean. NASA's Dawn mission entered orbit around Ceres in 2015, after several years orbiting Vesta. In 2017, the Dawn science team estimated that 10 percent of Ceres's mass is water ice, and they also detected an organic compound called tholins, prebiotic materials which may have facilitated the creation of life on Earth.[14] An ice-boring mission to Ceres could be a stepping-stone to more ambitious missions to the icy moons of Jupiter and Saturn, which are more promising locations for life but two to four times more distant from Earth than Ceres is. A human mission to Ceres is only marginally more difficult than one to Mars: farther to go but easier to land on, all within the scope of current technology.

As an industrial mining hub, Ceres may be in a good location, situated in the heart of the asteroid belt. There's plenty of water to be mined on Ceres, and that bit of gravity may be necessary to process ore from those other asteroids. From a spaceflight perspective, landing on and leaving Ceres would not be difficult, because of the lack of a deep gravity well. The little gravity there is would enable you to hop-walk on Ceres and set up a permanent industrial base. The other asteroids provide such a low-gravity environment that, from some of them, you could jump off into space. Ceres is certainly easy to track, too, so you will always know where to find it.

The question is, would you ever need such a hub? Asteroid mining will start with near-earth asteroids, because they are, well, near to Earth. Mining them down to nearly nothing and moving them into cislunar orbit might very well be a good thing for the long-term safety of the Earth. Whether we need the asteroids in the main belt depends on how much we plan to build in space, and serious construction projects—orbital cities and the like—are a century away.

The bottom line is that there seems to be no practical reason for living long-term on Ceres anytime in the next hundred years, when living on the Moon is easier. The technology allowing us to be out near Ceres even contemplating a settlement there, for economic reasons, would likely allow for the creation of a massive orbiting sphere with warmth and artificial gravity, which sounds far more civilized.

Living on an asteroid seems impractical, but how about living *in* an asteroid. Some asteroids are indeed city-size, with the mass and diameter of a mountain. You could hollow these out, create a subterranean space large enough for tens of thousands of people, and spin it to generate the centrifugal force that can simulate gravity. A chamber 100 meters in diameter spinning about four revolutions per minute would create an artificial gravity of 1G.[15] This would be different than the spin you experience on Earth with the Earth's rotation. You would be, from an outsider's perspective, underground and upside-down, with your head toward the core and your feet toward the surface, like water pinned inside a rotating bucket.

Austrian architect and civil engineer Werner Grandl estimates that, to maintain stability, an asteroid must have a minimum bulk density in its core of approximately 3 g/cm^3, close to the Earth's upper mantle, and many asteroids meet this criterion.[16] Asteroid homesteading—that's what this is. This would be your own world, with enough resources to support you and your family and community. Vertical farms could be closer to the surface, in non-spinning chambers, fed by filtered, reflected sunlight and artificial light. C-type asteroids can be upward of 10 percent ice, so there should be enough potable water and oxygen, recycled and exchanged with the CO_2 from the farms. You could 3-D print most of your tools, too. You could take your spacecraft to the big "trading post" at Ceres or Vesta to pick up supplies, akin to taking the wagon into town hundreds of years ago. Or some clever company based on Ceres could send you supplies via a drone.

Between the concepts of living in asteroids or using the resources to build orbiting cities, asteroids can provide enough material in the form of water, air, fuel, metal, soil, and nutrients to support ten to a hundred trillion humans.[17] Let's propel this concept even further: should hydrogen fusion become available, you can turn your mountain-size asteroid into a well-protected space ark, set off for a distant star system, and reach it in a few hundred generations.

My prediction: Asteroid mining by the 2030s, entirely robotic; robotic probes to Ceres and Vesta by the 2040s; boots on Ceres by the 2060s; small, semipermanent presence on Ceres by the end of the twenty-first century; asteroid homesteading in the twenty-third century.

6

Living on Mars

Everyone's excited about Mars for good reason: if permanent settlements are possible on any natural body in the Solar System in this century other than on Earth, they will be on the red planet. By "settlements," I mean a place where adults can nurture children and where new cultures can develop. Massive orbiting cities within the Earth–Moon system are certainly feasible, but because of practical reasons—the expense and complexity to build them and the will to live in them, when they don't offer anything the Earth doesn't—don't expect this to happen anytime soon. How does Mars compare with other options? The Moon is cold and stark, draped by a black sky, and has the low-gravity problem—fine for mining, fine for science excursions, fine to visit, but not fine for raising children. Venus solves the gravity problem, but its surface is hot enough to melt lead, and it's hard to imagine the lure of going all the way there to live in blimps, or floating cities, high in the Venusian atmosphere. Mercury's gravity is similar to Mars's, but you still have that high temperature problem.

So, no other solid body in the Solar System other than Mars offers possibly suitable gravity and a reasonable temperature. Mars actually seems tameable. It's familiar. Its mountains, canyons, gorges, and valleys call to us. The majestic Olympus Mons is nearly three times taller than Mount Everest; the stunning Valles Marineris is a canyon as long as the United States is wide. Also, Mars has every chemical element we need to survive. It once was warm and wet enough to harbor life, and it could be that way again if we dedicate ourselves to the task. The million-dollar question is, though, can Mars sustain a natural human settlement? Can the now-arid valleys be turned into rich loam, and can the dry riverbeds flow again—if, not with water, then with some kind of promise of a fruitful life to attract and keep generation upon generation? Can lonely outposts grow into towns and then cities with vibrant biological and economical ecosystems?

The first step in sending humans to Mars, of course, is having a plan to send humans to Mars. For the United States that means a plan that doesn't change every four to eight years with each new administration. China has a plan to establish human colonies on Mars by the 2040s. And you can bet your bottom yuan that they aren't going to change that plan if President X is replaced by President Y. Plans may be delayed, but they likely won't be outright canceled for reasons as trivial as a change in administration. NASA has yet another new plan, more poetry than roadmap, describing intangible stages that move us from "earth reliant" to "proving grounds" to "earth independent" through a series of technological goals, not missions per se.[1] It's a poem that weaves all of NASA's disparate, unfunded ideas into an interconnected timeline without actual dates, rather than the concrete demand for deliverables that characterized the Apollo era.

The problem with US efforts in achieving milestones in human spaceflight is that NASA has to contend with a concept known as "the president's vision for exploration."[2] All politics aside, who

Mars. This mosaic image comprises 102 Viking Orbiter photographs. Featured prominently is the Valles Marineris canyon system, more than 2,000 kilometers long and up to 8 kilometers deep. The western edge may be a scenic place to visit or live, with the three Tharsis volcanoes (dark spots), each about 25 kilometers high, visible to the far west.

really cares what any president thinks when it comes to space? It doesn't matter. Give NASA the money and let its scientists and engineers decide where and how to go. For reasons stemming from the militaristic origins of NASA, however, Congress and the US president micromanage NASA human space activities and can cancel programs on a whim or, often worse, direct them to be done in certain congressional districts, regardless of the inefficiency. This is why both the space shuttle and ISS projects were such boondoggles, with cost-plus contracting and an ineffective supply chain scattered across the country that all but guaranteed being over budget—transforming NASA into more a constituency-driven agency than a mission-driven agency.

Indeed, NASA excels brilliantly when it is mission driven, as is the case for all space science missions. NASA has launched every probe that has visited Jupiter, Saturn, Uranus, Neptune, and Pluto, all gifts to humanity; and the United States dominates space-based, multi-wavelength astronomy with telescopes the likes of Hubble, Chandra, Kepler, and WMAP. How this works is simple: astronomers propose a mission and provide a sound case for the science and ability to build it; NASA chooses a certain number of missions each year from these proposals; and the astronomers go about building them, almost always within budget and on time. NASA and the United States are the envy of space agencies worldwide in this regard. But when astronauts are added to the mix, in a big project conceived by US politicians, things quickly go awry.

I and others would argue that the primary reason NASA has done so little in *human* space exploration for the last forty years is that the space agency is directed by ever-changing US presidents (twelve since its creation) and micromanaged by the US Congress. The "vision"—stay in low-earth orbit; no, go to the Moon; no, go to Mars instead; no, go to the Moon—is a blurry, moving target. Jimmy Carter pushed for space science over human activities; Ronald Reagan supported the ISS as a stepping-stone to a larger presence in orbit; George H. W. Bush pushed for a return to the Moon and then a journey to Mars; Bill Clinton focused on cooperation with the Russians to complete the ISS, which was over budget and less international when he took office; George W. Bush wanted to return to the Moon; and Barack Obama wanted to skip the Moon and go to the asteroids and Mars.[3] Donald Trump has advocated, at various times, to go to the Moon, or to Mars, and to create a Pentagon-led space force.[4]

Had Barry Goldwater beaten Lyndon B. Johnson in the 1964 US presidential election, the United States would not have landed on the Moon. Goldwater was outspoken in his criticism in the Apollo mission and said the civilian space program was draining too much money from military space efforts.[5] But Johnson won; and the planned July 1969 Apollo 11 mission was so close to reality that Richard Nixon couldn't cancel it when he took office in January 1969. Nixon would wait three more years before ending Apollo prematurely.

NASA has an annual budget of about $20 billion and could likely get to the Moon or to Mars or both with that money, if that were all NASA had to do. But there are other major programs in rocketry, space science, aeronautics, Earth science, and human spaceflight on NASA's agenda. More problematic, the agency is stuck with its annual large financial commitment to the ISS, a fifth of its budget. In the 1990s, engineers David Baker and the aforementioned Robert Zubrin (author of *The Case for Mars*) from the Martin Marietta Corporation developed a mission concept called Mars Direct that they claimed could establish a permanent human presence on Mars in a little over a decade for about $40 billion in today's dollars, or a few billion dollars per year over the course of its development, and then about $4 billion yearly to sustain.[6] That's equivalent to the current annual ISS budget. Alas, although Mars Direct received serious consideration by NASA and was met with praise by many within the organization, the vision hasn't yet matched any congressional or NASA directive.

The bottom line is that permanent Martian settlements are within reach from fiscal, biological, and engineering perspectives. Here's how this might play out over the next two decades.

The Voyage Begins on Earth

The landscape of Mars is similar to the Antarctica landscape: cold, monochromatic, and remote, but with captivating beauty. We aren't living in Antarctica in large numbers because of the challenges of life there, particularly during the sunless winter. Living on Mars in large numbers would be even more challenging for lack of a creature comfort we call air. That's air to breath, air to form enough atmospheric pressure to keep our blood cells from popping, and air to block harmful radiation from the Sun and beyond. There's far more air on Mars than on our utterly airless Moon, but the air is thin nonetheless and comprises almost entirely carbon dioxide, with about a percent or two of nitrogen and argon and a whiff of oxygen. We can't breathe that and survive. Some Earth organisms might be able to, if only it weren't so cold there.

On Earth, the atmosphere is about 78 percent nitrogen, 20 percent oxygen, and 1 percent argon, with traces of carbon dioxide and other gases. The air pressure at sea level is about 100,000 pascals, neatly called 1 atmosphere. On Mars, the air pressure is 600 pascals, about 0.6 percent of an atmosphere. This is the main challenge on Mars: nothing breathable and nothing thick enough to offer protection from solar and cosmic radiation or from the liquids and gases in your body expanding. Almost everything that follows about settling on Mars concerns overcoming the atmosphere issue. We can only hope that there is no gravity issue, that 0.38G is enough for a fetus to develop and a child to grow up healthy. Barring an experiment on the ISS to test simulated Martian gravity, which NASA more or less has barred, we won't know how well we'll fare in 0.38G until we get there and try it.

This, by the way, is the difference between settling on Mars and settling on the Moon. The Moon has both the atmosphere *and* gravity issues. Mars may have only the atmosphere issue.

Can living first on the Moon teach us anything about living on Mars? Absolutely. A basic shelter would be the same on both bodies: a habitat underground or otherwise well shielded from radiation, with adequate pressure, air, food, and water. The key difference is in the long-term strategies for inhabiting these places. Life on the Moon, partly because of the Moon's proximity to Earth and partly because of the brutal temperature extremes in its twenty-eight day–night cycle, will be dependent on the Earth for centuries to come. Initial skill sets will include coping with the low gravity and mining for resources such as ice, which is thoroughly mixed with the Moon regolith and in low concentrations. Mars, however, partly because of its *distance* from Earth and partly because of its relatively manageable temperature and day–night cycle, will be approached as a frontier that ultimately, and quickly, requires a high level of self-sufficiency for this planet to be an economically viable human destination. On Mars, skill sets would include making rocket fuel by converting carbon dioxide into methane and oxygen, as well as building vast greenhouses for growing food in natural sunlight.

We could learn how to live in space shelters by first going to the Moon. Knowledge would be gained in terms of transportation, supplies, storage, and suiting up to go outside. This in no way would be a waste of time, as some in the "Mars or bust" crowd might argue. Everything we do in space teaches us something about living in space, albeit at a price. Moon habitats on the lunar surface would be the "real thing," with real danger and real-life situations that need to be remedied in real time, with help from or evacuation to Earth three days away. In this regard, a Moon habitat provides experience for Mars (six to nine *months* away) that can't be provided in a simulation on a submarine or Earth-based analog habitat. That said, Earth-based analog habitats are relatively cheap, and they serve as baby steps to both the Moon and Mars.

Desert Hideaways

I discussed HERA (Human Exploration Research Analog) and HI-SEAS (Hawai'i Space Exploration Analog and Simulation) in Chapter 1. The biggest, if not most obvious, limitation to these analogs is that these aren't real space adventures. NASA very well could be merely studying the psychological stresses of prison. On a true trip to Mars, you are going to Mars. That's a rather massive psychological lift that could easily trump any minor inconvenience about cramped quarters for a year or two. I personally think the psychological risks are overblown. In the first missions to Mars, there will be a tremendous sense of purpose. In subsequent human migrations with ordinary people, technology will have advanced to enable shorter trips on more comfortable spacecraft.

Other Mars analog missions on Earth focus on living, working, and moving about on Mars, which provides good practice. The Flashline Mars Arctic Research Station (FMARS) and the Mars Desert Research Station (MDRS) are two surface-exploration habitats operated by the Mars Society, a nonprofit space-advocacy group founded by Zubrin and others in 1998. These facilities look like what you might have in mind when you think about the first Mars habitats: simple cylindrical structures with an airlock, some antennae, and a vehicle parked outside. FMARS is on Devon Island in Baffin Bay in the Canadian region of Nunavut, a polar desert ecology similar in many ways to the Martian polar regions. MDRS, in a desert environ of southern Utah, is similar to FMARS in design but with a greenhouse and astronomy observatory. With these two facilities and a limited budget, the Mars Society works out concepts presented in Zubrin's book *The Case for Mars* by going through the routines of growing some food and surveying and excavating the land nearby, discussed later in this chapter. The locations were chosen for their similarity to Martian geology. Participants are volunteers, many of whom are college students. The MDRS got a taste of "real" problems when its greenhouse was destroyed twice, first by wind and then by fire.

NASA has its own desert hideaways. NASA's Desert Research and Technology Studies (Desert RATS) in Arizona comprises a series of annual field trials testing technologies for living and moving about the surface of the Moon or Mars, a revival of the Apollo mission tests from the 1960s. In partnership with universities, NASA tests vehicles, suits, tools, and more. This includes the All-Terrain Hex-Limbed Extra-Terrestrial Explorer, which looks like it was lifted out of the fictional world of *Star Wars,* a six-wheeled spider-shaped vehicle capable (we hope) of navigating any kind of lunar or Martian terrain. NASA has tested another robotic device to flatten out a stretch of soil and sinter it to create a landing and launch pad.

The NASA-funded Haughton Mars Project (HMP) is close to FMARS on Devon Island, only larger. HMP is situated in the Haughton Crater, a Mars-like landscape of dry, unvegetated, loose rocky terrain. This makes the HMP a good analog for testing robotic mining techniques that would be used on Mars. As with FMARS, the remoteness and lack of infrastructure allows for the testing of energy storage, communications, and telemedicine.

HMP is the brainchild of Pascal Lee, a NASA engineer who co-founded the Mars Society as well as the international Mars Institute. Between Desert RATS and HMP, NASA and its partners are testing the basic necessities that seem to be part of the backdrop of Mars-based science fiction books and movies: pressurized rovers, robotic diggers, and flexible pressurized suits.

Living Off the Land

Other key projects include in situ resource utilization (ISRU), gadgetry to help "planetary pioneers" live off the land. Temporary and certainly permanent lunar and Mars settlements will depend on ISRU, because the physics of the rocket equation essentially prohibits us from bringing superfluous and heavy materials such as air and water. And, yes, air is heavy. We breathe 550 liters of oxygen each day, or about a kilogram. Packing enough oxygen for a four-member crew to Mars for two-year mission to Mars would add three metric tons (6,600 pounds, $30 million) to the journey.

The Regolith and Environment Science and Oxygen and Lunar Volatile Extraction (RESOLVE), developed by NASA and the Canadian Space Agency, is part rover, part drilling platform, part oven, and part laboratory, designed to produce water and oxygen gas. This mobile unit can extract a one-meter geologic core, pulverize it, heat the oxygen-rich rock to 900°C, liberate the oxygen, mix it with hydrogen, and produce water in about an hour.[7] Getting water from a rock, just like Moses. Designed for the Moon, this instrument could be used on Mars to find other important volatiles such as ammonia, carbon monoxide, helium, and hydrogen. Related to this, the Precursor In-situ Lunar Oxygen Testbed (PILOT) demonstrated the ability to extract from regolith oxygen that was locked away in minerals, and this might have some use on Mars.[8] Mars has more water in its regolith than the Moon, though, and it would be less energy intensive to extract oxygen from the water.

On Mars, there's O_2 in the air, but it's tied to CO_2. The Mars Oxygen In-Situ Resources Utilization Experiment, or MOXIE, can pull in CO_2, heat it to about 800°C, pass it through a catalyst called a solid oxide electrolyzer cell, and produce breathable oxygen via the equation $2CO_2 \rightarrow 2CO + O_2$. The poisonous carbon monoxide by-product could be used for fuel or further converted to methane. The MOXIE instrument, fully tested in the lab, is planned for NASA's Mars 2020 rover mission. If it works, NASA is prepared to scale it up a hundredfold. The MOXIE prototype, running on only 300 watts to produce 10 grams of O_2 per hour, can be solar powered. A full-scale version for astronauts on Mars likely would require a radioisotope thermoelectric generator, which converts the heat of radioactive decay into electricity. This is the energy source powering the electronics of most spacecraft sent beyond Mars, where solar energy is too weak.

Oxygen is needed not just to breathe but also to burn. Hydrogen, methane (CH_4), and all carbon-based fuels cannot ignite without the presence of oxygen, the oxidizer. So, while greenhouses are nice in creating a perfect CO_2 to O_2 ratio for plants and animals, we still need more oxygen for our burning needs, so to speak. MOXIE instruments could be set up in great numbers to create large stores of oxygen.* You also might see the beauty of chemical cycles. Mars has everything we need for a civilization, just not in the form we want it. But hydrogen-, carbon- and oxygen-based

*Storing gases is a more complicated issue, requiring heavy tanks, meaning more mass on a spacecraft unless they can be manufactured on Mars from local materials.

molecules are all interchangeable, given an energy input. Nothing need be wasted if used on Mars. Liberated hydrogen from water goes back to water when burned as a fuel with oxygen. CO_2 coverts to O_2 and CO, and CO goes back to CO_2 when burned with oxygen. $CH_4 + 2O_2 \rightarrow CO_2 + 2H_2O$. Matter cannot be created or destroyed. The only time that resources are lost to us is when they can't be practically collected, such as during a rocket launch.

China, I should note, has embarked on several of its own ISRU projects and desert field studies. China is constructing a multimillion-acre Mars world in its Qaidam Basin, an arid region on the Qinghai-Tibet Plateau with barren landscape resembling geographical conditions found on Mars. The complex will be mostly scientific but also will include a Mars-themed tourist attraction with exploration camps and simulated Martian experiences.[9] There are dozens of other nifty projects worldwide—in Europe, Japan, and Dubai, a newcomer to the space game—all with clever acronyms and all fairly far along in testing. It seems that we're ready to go to Mars. Well, almost ready.

First Obstacle Is Leaving Earth

In July 1989, on the twentieth anniversary of the Apollo 11 Moon landing, US president George H. W. Bush delivered a speech atop the steps of the National Air and Space Museum in Washington, DC, proposing a comprehensive plan for human exploration of space that included construction of an orbital space station (then called Space Station *Freedom*), a permanent return to the Moon, and a human mission to Mars within thirty years, target 2019, a half century after the first steps on the Moon. This was called the Space Exploration Initiative (SEI). NASA embraced the idea and commissioned the now infamous 90-Day Study, which ninety days later calculated the costs to be $500 billion—more than $1 trillion in today's dollars.

Sadly, this wasn't so much space policy as it was space entertainment, a delicious plan with little basis in reality. Part of the problem with the SEI is that it relied on the then-unbuilt Space Station *Freedom,* the original ISS concept that would soon go so far over budget that the word "Freedom" was replaced by "International" to get other countries to help pay for it. The plan also seemed to include everyone's pet projects, with fuel depots to be added to the space station, bases placed on the Moon, and a massive ship built on the Moon to fly to Mars, only to have a portion of the crew land for just a few days to plant a US flag. Bush had created a National Space Council in 1989 to guide the initiative, to be led by his vice president, Dan Quayle, another critical mistake. Members of the Space Council were at odds with NASA's lack of initiative in finding innovative solutions to expensive technological problems, and Quayle never had the savvy to intervene. NASA itself incorrectly assumed that Bush's proclamation would translate into a funding windfall. Neither the White House nor NASA engaged the US Congress in discussions about the plan. Congress saw the price, the most expensive project since World War II, and got sticker shock.

This enormous price tag was coupled with a growing belief among many Americans that NASA no longer had the "right stuff" to pull off something so grand. The space shuttles weren't living up to their promise, and the Space Shuttle *Challenger* had exploded upon launch in 1986, killing all seven crew members, including a schoolteacher. Within a year of Bush's SEI speech, the multibillion-dollar Hubble Space Telescope would launch with a flawed mirror, a bonehead error. This also was post-Reagan America: a struggling economy with rising deficits and the politico-fiscal realities of the day killed the Space Exploration Initiative within three years.[10] NASA historian Thor Hogan sums up the SEI this way: "In the final analysis, the demise of SEI was a classic example of a defective decision-making process—one that lacked adequate high-level policy guidance, failed to address critical fiscal constraints, developed inadequate programmatic alternatives, and garnered no congressional support."[11] Hmmm, this sounds to many like the 2019 call for a human Moon landing by 2024.

One positive step in this era is that NASA administrator Dan Goldin, appointed in 1992 after the Space Exploration Initiative meltdown, made studying Mars robotically a main objective in NASA's exploration of the Solar System. And that has become immune to presidential visions and cancellations. Also during that era rose the irascible Robert Zubrin. He saw a way to get to Mars at a twentieth of the price posed by the SEI, using existing technologies. His plan was called Mars Direct; and although NASA ultimately rejected it for safety reasons—too much uncertainty, a risk-averse NASA thought—that plan posed circa 1995 eerily resembles NASA's ideas today, which may ultimately get humans to Mars.

Zubrin envisioned tremendous savings on a Mars mission by going lean and living off the land, as opposed to the approach NASA was considering in the Space Exploration Initiative, which packed everything needed for the round-trip voyage: fuel, air, and water. All three of these resources are present on Mars. With the Mars Direct plan, we first send an unmanned Earth Return Vehicle (ERV) and a cargo of hydrogen gas, which arrives in about six months. Upon landing, a rover rolls out with a machine that uses the hydrogen to covert the carbon dioxide (CO_2) in the Martian atmosphere into methane fuel (CH_4) and water (H_2O).* The water

*Zubrin tweaked this plan with the more recent discovery of abundant water on Mars, which eliminates the need to send hydrogen. Instead, we extract hydrogen from that water, possibly underfoot.

can then be broken into oxygen and hydrogen. Over the course of a year, these two reactions will produce enough fuel and oxidizer to propel the ERV back to Earth, along with extra fuel to power ground vehicles and extra water to drink or "breathe," once split into hydrogen and oxygen. With 6 tons of hydrogen from Earth, we could make 108 tons of methane and oxygen on Mars.[12] The chemistry is rather straightforward. In fact, Zubrin built such a device for making methane with a small amount of NASA funding, to demonstrate the feasibility. The challenge would be to land this safely and have it work in a cold, low-gravity, low-pressure environment.

Note, Earth and Mars are at their closest point about every twenty-six months. So, about two years after the ERV launch, we send two more rockets to Mars: another ERV (to start another cycle of fuel and water creation) and a habitat module with a four-person crew. The crew lands lightly without the mass of fuel, air, and water weighing them down; and the first ERV is there for their return, simply waiting to be loaded up with fuel already made on Mars. The second ERV serves as an emergency return ship in case the first one is somehow damaged. The crew returns after about 1.5 years on Mars, as the planets draw close again. Meanwhile, the cycle repeats. Four more people arrive at Mars with an ERV and another habitat. Every two years brings another crew, ERV, and habitat; and slowly a Mars village develops as habitats are joined together via underground passages.

In the Mars Direct plan, every two years could bring new equipment, too, such as a small nuclear reactor and solar panels for energy. Over the years, the crew would have enough tools to start building a Martian infrastructure, with small factories churning out plastics, glass, and ceramics.

NASA's main problem with the Mars Direct plan was that the agency didn't trust Zubrin's numbers, which it felt were too optimistic in terms of fuel needed and mass not needed. Fair enough. NASA tweaked the plan with a new design that would send three spacecraft at a time, which Zubrin dubbed, not entirely pejoratively, Mars Semi-Direct. The cost was calculated at $55 billion spread over ten years. This would have fit into NASA's budget while allowing it to pursue other activities, such as space and earth science.

Mars Direct had a few other issues, too, such as sheltering the ERV for two years from the intense UV and persistent dust. Little things like this add dollars to missions. In the end, NASA never green-lighted Mars Direct or Semi-Direct. Some Mars-mission advocates suspect that the agency didn't want an outsider to tell it what to do or how to do it. Robert Zubrin and David Baker, his colleague who helped develop this scheme, were just a couple of regular guys, after all, not even part of NASA. NASA focused instead on the expensive task of building the ISS.

Second Obstacle Is Remaining Healthy While Traveling There

OK, let's pretend we have the money, the rocket, and the will to get to Mars. This winning combo likely will arise from some government, either China and its partners or the United States and its partners, whichever those partners might be in the 2030s. And yet, some billionaire might very well fund an excursion if it is cheap enough. Regardless of the funding and motivation, the question must be, how do we make the voyage? Traveling to Mars is more difficult than *living* on Mars.

As discussed in Chapter 2, the health risks associated with a trip to Mars are serious. I think low gravity is the gravest danger to mission success, so I'll elaborate on that issue here.

The travel time to Mars is six to nine months. If that time is spent in a microgravity environment, bones and muscles will grow weak. Astronauts who spend six months on the ISS cannot function well when they return to Earth, even with their two-hour exercise regimen while in orbit. They often cannot walk at all for at least a day and suffer from nausea upon their return. That's not too bad when you have a team on Earth to lift you out of the spacecraft and carry you to a wheelchair. There will be no such welcoming committee on Mars. Astronauts will need full control of their bodies—immediately. More than that, the prolonged microgravity has shrunk their muscles, lessened their bone density at a rate of at least 1 percent per month, and damaged their vision. The only comfort in landing on Mars instead of on Earth after nine months in space is that Mars is 0.38G, so the effects of terrestrial gravity might not be as extreme.*

Recall, NASA thoroughly tested the capabilities of astronaut Scott Kelly after his 340-day stay on the ISS. He was barely able to walk upon landing but got better day by day. He could perform many mechanical operations, but his coordination was clearly off. There's a name for this: Space Adaptation Syndrome (SAS). The brain gets confused when it doesn't know where "down" is. SAS hits astronauts when they arrive on the ISS and then when they return. It's not so bad going up, because you have crewmates on the ISS already adapted. On a mission to Mars, mission controllers on Earth can help the crew for the first couple of days. But on Mars, with these exact symptoms that Scott Kelly experienced, the mission could be in peril.

*One health test could be landing an astronaut on the Moon after six months in space, to see how he or she does in 0.16G, but I have not heard of any discussion about this possibility.

Performing resistance training for hours a day during the trip will help. So, there are basically two ways to go to Mars: travel in microgravity, exercise along the way, and hope that in the 0.38G environment the SAS on Mars will be only 38 percent as severe; or spend the money to rotate the spacecraft to create artificial gravity. It's my take that gravity is not a luxury. Sending astronauts to Mars with no artificial gravity will leave them unable to perform their mission, tantamount to a death sentence.

In many science fiction movies, the gravity issue is just ignored. In *Star Trek* and *Star Wars,* there's an assumed gravitational field generator that provides the necessary gravity on the spaceship, a technology that doesn't exist and is in the realm of warp drive, beyond the confines of known physics. On a trip to Mars, we'd need to induce an artificial gravity by creating centrifugal force. A simple design advocated by many engineers, including Zubrin in *The Case for Mars,* is a capsule tethered to a counterweight. This would fully deploy in space as the two sections detach and the cable joining them unwinds to set the sections apart by 1,500 meters. Then engines are fired once to send one end tumbling over the other, all the way to Mars. One revolution per minute would generate a Mars-like gravity; nearly two rpm would generate an Earth-like gravity. Maybe not graceful, but effective. A more elegant design is described in Andy Weir's science fiction novel *The Martian.* The spacecraft is called the *Hermes,* a hundred-meter-long tube with a central rotating hub that's shaped like a Ferris wheel. Mars-like gravity is felt in the outer parts of the wheel, where the crew spends most of their time.

The fictional *Hermes,* so large that it is assembled in space, is far beyond anything NASA is considering. Sadly, NASA doesn't appear to be considering the simpler tethered design. NASA's current plan is to send a four-person crew to Mars in a spacecraft now under development called *Orion,* launched by the SLS. *Orion* doesn't rotate. NASA is relying on the ISS-developed exercise regime to keep the astronauts healthy, uncertain of the feasibility of artificial gravity and its added cost to the mission. The notion of saving money is silly, actually, when you consider the added mass of exercise machines and the extra food needed to compensate for extra expended calories, let alone the wasted time exercising for hours a day.

NASA is also assuming that astronauts would be able to recover on their own once on Mars with simply a day or two of rest. Although it is true that astronauts can partially recover in a few days when returning to Earth, they are recovering under the care of medical professionals. Coordination issues, muscle weakness—maybe these are manageable without medical care. But one other symptom that occurs when microgravity meets macrogravity is orthostatic hypotension, dangerously low blood pressure caused by a combination of dehydration and cardiovascular deconditioning. The heart has difficulty pumping blood to the brain, causing returning astronauts to feel faint and occasionally black out. Also, the astronauts' brittle bones leave them susceptible to fractures from everyday activities, such as walking or lifting, which they will be doing wearing heavy equipment. Their mission begins at the point when most long-duration missions in microgravity end.

Artificial gravity can work—it *is* tested and can be further tested. Japanese scientists placed a rotating contraption in their own part of the ISS, to house mice. The mice lived for thirty-five days on the ISS in artificial 1G and had absolutely none of the adverse effects of living in orbit compared with mice living in microgravity for the same period.[13] A more ambitious project is the Multigenerational Independent Colony for Extraterrestrial

Habitation, Autonomy, and Behavior Health (MICEHAB), a mission that would send rodents into space by themselves, with robotic caretakers, to see how well they reproduce in artificial gravity set at 1G or 0.38G. MICEHAB is only in the design-concept phase, though, with no funding. MICEHAB would be an ideal way to test the feasibility of human gestation on Mars.

I also described the radiation risk during a trip to Mars in Chapter 2. It comes down to acceptable risk, at least for the initial sets of human missions. To transport the multitudes to Mars to create settlements, though, we need to figure out how to effectively shield from radiation. If you were caught in a solar storm while in a spacecraft traveling to Mars, you'd get a huge dose of nearly 40 rem, or 40,000 millirems, equivalent to forty full-body CT scans. A simple storm shelter on the spacecraft could lower this to five rem, bad but not horrible. A tiny, well-shielded room, such as a food pantry behind tanks of water, to use for a few hours until the storm passes could do the trick.

Apollo astronauts picked up about 1 rem over the course of about ten days. Among the twelve astronauts, one died at age sixty-one (heart attack); one died at age sixty-nine (motorcycle accident); one died at age seventy-four (leukemia); and the others all made it into their eighties, longer than the average US male life expectancy. Although the sample size is small, we can be somewhat confident that the Apollo missions didn't shorten lives. Scott Kelly picked up 8 rem during his year on the ISS, which is 3 rems over the allowable limit for US workers. Only time will tell whether Kelly will develop cancer; and even then, he represents a sample size of one. Did the radiation cause it? One in three people will die of cancer. Astronauts on Skylab received a full-body dose of 17.8 rem in just two months.[14] Alan Bean was one of them. He also went to the Moon. He died at age eighty-six of a sudden, undisclosed illness. NASA estimates that cosmonauts aboard Mir for a year received 21.6 rem.[15] No word on their health; the Russians are tight-lipped. Travel to Mars ups the ante. The radiation exposure on the trip to Mars is far higher than any US "worker" takes, aside from being in a nuclear accident.

Third Obstacle Is Landing Alive

Traveling to Mars is complicated by one more issue—landing! No space agency yet understands how to land so much mass on Mars. The heaviest objects to (almost) land on Mars were the Soviet Mars 2 and Mars 3 probes, each 1,210 kilograms. Mars 2 crashed on the surface, and Mars 3 failed a few seconds after landing, although dust storms at the time didn't help. The US Viking landers were half that weight. Probes got even lighter until NASA's 900 kg Mars Science Laboratory, with the *Curiosity* rover, in 2012. Plans for Mars habitation require several tons of cargo to be landed. The thin Martian atmosphere heats incoming vehicles and yet limits the amount of aerobraking possible via parachutes, the worst of both worlds. There's no ocean for a splashdown, either. NASA developed the so-called sky crane to lower *Curiosity,* essentially a device to fire engines in reverse to slow the descent. Failure to land a robotic probe (and there have been many) is a loss of a few billion dollars. Failure to land a habitat containing astronauts is a loss of life—and billions of dreams. The flawless landing of the NASA InSight probe in November 2018 bodes well. The probe entered the Martian atmosphere with about 600 kilograms, some of which was fuel; and it landed at an elevated area, thus with less atmosphere to use for braking.

Martian Highways

What's one good way to beat the radiation and microgravity problem? Get to Mars as fast as you can. In theory, we could get there in six to nine days with modern technology (nuclear bombs). In practicality, the trip is between six to nine months. Here's why.

The Moon maintains a relatively consistent distance from Earth, about 384,000 kilometers on average. At perigee, or its closest, it is 363,000 kilometers away; at apogee, or its farthest, it is 405,000 kilometers. So, it doesn't matter when we choose to travel to the Moon, at perigee or apogee, because the difference in rocket travel time would be only a few hours. The distance to Mars varies greatly, however. Sometimes the red planet is on the opposite side of the Sun. Mars can be as close as fifty-five million kilometers and as far as four hundred million kilometers. Timing is everything. You want to leave Earth at just the right point to hit Mars, a moving target, as it approaches closer to us. This launch window comes about every twenty-six months.

But three other primary factors are at play dictating how fast we can travel: g-forces, fuel efficiency, and the need to slow down. Slowing down is related to delta-v, changing velocity to go from one kind of orbit to another, which requires firing the engines in the opposite direction. If you don't slow down to the precise speed needed, you overshoot your target and will have to expend even more fuel, if you have it, to get back. The Apollo missions took about three days to get to the Moon. This somewhat leisurely pace allowed the astronauts to easily slow down and enter into lunar orbit. The slower you are moving, the easier it is to slow down. NASA's New Horizons probe to Pluto didn't need to stop at the Moon; it zipped past the Moon in just eight hours, thirty-five minutes, traveling 58,000 km/hr. At that speed, it would have passed Mars in about forty-one days, had Mars been at its closest point to us. So, in theory, with modern rocketry and chemical fuel, if the right launch window is open, we could fire a rocket at Mars with supplies, and it would crash onto the Martian surface in forty-one days. Sloppy and destructive, but fast.

Fuel efficiency prevents us from taking the very shortest path, though. New Horizons was traveling at the escape velocity needed to get to the edge of the Solar System. In theory, you could shoot off Earth at this velocity in a straight line on a direct path to where you expect Mars to be when you get there and then, halfway, fire the engines full blast to slow down, getting you there in sixty to eighty days. But this would require so much fuel as to make it impractical to try it, relating back to the tyranny of the rocket equation, discussed in Chapter 3, and not being able to lift off the Earth with all that fuel. The concept becomes approachable if we refuel in low-earth orbit, but a lot of chemical fuel is still needed. So, instead, rocket engineers send spacecraft to Mars via the Hohmann transfer orbit, a swirling elliptical orbital path that gradually takes the spacecraft halfway around the Sun as it inches away from Earth's orbit and toward Mars's orbit. This is one of the most fuel-efficient ways to get to Mars and, for now, fuel efficiency dictates everything. It takes about nine months. There are more direct ways, but these would be analogous to going against the flow of water. The Hohmann transfer has essentially been our highway to Mars since the late 1960s; most probes have taken this road.

Zubrin, in his Mars Direct plan, suggests a path similar to the Hohmann transfer but at a slightly different trajectory, requiring more fuel but arriving to Mars a little faster, in eight months. An even faster path to Mars would be an Earth–Mars cycler, a vehicle moving in a very specific trajectory that never lands but rather orbits the Sun in a path that takes it close to Mars and then close to

Earth over and over. Think of it as a space train. This path is called a free-return trajectory because the gravity from one large body (e.g., Mars) whips the vehicle back to the other body (Earth) in perpetuity, with little need for fuel. It stays in orbit, propelled by gravitational slingshots. Shuttles are needed to get to it as it passes by Earth or Mars. Of the many available flavors of free-return trajectory, one under consideration is called the S1L1 trajectory, which would take about 150 days, or five months, to get to Mars.[16]

Zubrin's Mars Direct path is feasible with current technology, but the Earth–Mars cycler faces two challenges. The precision needed to achieve this trajectory and then dock with the cycler has never been accomplished. And, of course, we'd need to build a big spacecraft in orbit to serve as the cycler. Once we start going to Mars regularly, though, the Earth–Mars cycler would be a superior system. Some fuel, however minimal, would be needed to keep the cycler on proper trajectory and add any speed lost during the docking and detachment times. Buzz Aldrin has his own version of this, called the Aldrin Mars Cycler, and has said the trip can be done in six months. The added charm of the Earth–Mars cycler is that part of the cargo could be dropped off at the Martian moon Phobos, which will be an important base, as explained below.

Want to get to Mars even faster? Freeman Dyson's Project Orion, the nuclear propulsion rocket system, would have been powerful enough to get to Mars in about a week. That's not to be confused with the *Orion* spacecraft, NASA's current proposed vehicle for travel beyond low-earth orbit, a tiny thing for a four-person crew. Project Orion was a classified government project from 1957 to 1965 to build a 4,000-ton building-size spacecraft powered by thousands of nuclear bombs. Nuclear fuel is far more efficient than chemical rocket fuel, so a spacecraft would be able to carry the necessary fuel to accelerate to a great speed and then flip over halfway through to slow down. Project Orion, of course, never came to fruition; the design was crawling with safety issues (you know, atomic bombs), but the concept remains viable. But humans can't accelerate too quickly without encountering deadly g-forces, that force of being pinned to your seat during acceleration.

The fastest that humans have traveled was on the Apollo 10 mission around the Moon in May 1969, the so-called dress rehearsal for the Moon landing a few months later. The three-man crew peaked at 39,897 km / h on their return. Speed isn't the danger, per se. From a biological stance, humans could travel at near light-speed in a protected vessel if technology allowed. The issue is the rate of acceleration. We need to be eased in to these high velocities. It's just like driving on the highway: on the on-ramp, you accelerate from about 30 km / h to 90 km / h in a few seconds, a feeling of about 0.5G pressing against you.* Once you reach cruising speed, the g-force diminishes. But look at the trauma caused by fast braking, when the body goes from 90 km / h to 0 km / h in a few seconds, a rapid deceleration, even strapped in. In space, traveling a thousand times faster than in a car, acceleration to top speeds needs to occur over the course of several days, and then be repeated in reverse when the time comes for deceleration.

During takeoff, astronauts experience forces between 3G and 8G for a few minutes; but because they are strapped in and positioned on their backs, the g-force is in the chest-to-back direction and doesn't cause blackouts—that is, it doesn't force blood out of their brains and into their feet. Then they reach cruising velocity, and all is calm. On a Dyson-inspired trip to Mars, the acceleration

*A sports car accelerating from 0 to 60 mph in 2.74 seconds will produce a 1G force.

to such great speeds would keep you pinned to your seat and, without some kind of shock absorber, turn you to mush. The acceleration force of an atomic bomb is in the range of 10,000G. On paper, Dyson figured he could cushion the acceleration to about 4G. Efficiencies in travel times would come with longer trips, such as to Alpha Centauri, where you have space in which to slowly accelerate. Dyson calculated that a nuclear engine could send a hundred-ton spacecraft with eight astronauts to Mars at a comfortable g-force level in about sixty days, with half the weight being actual payload.[17]

The other issue about going too fast, which will pose a problem for interstellar travel when we exceed 10 percent the speed of light, is the cosmic debris striking the spacecraft. Think of bugs on the car windshield. Or pebbles. The higher the speed, the easier for that little pebble to crack your windshield. At 10 percent the speed of light, microscopic dust particles in space would tear through the spacecraft like bullets through paper. There are no real solutions to this; the problem is so far off that it isn't considered worth tackling now.

Mars Plans: Peculiar, Practical, and In Between

Among the more peculiar ideas for settling the red planet was Mars One, a private project run by a small Dutch company. These were the details: send an unspecified number of ordinary people on a one-way trip to Mars at an unspecified date to live in an unspecified habitat, all funded by an unspecified revenue stream that might have included crowdfunding and selling the rights to documentaries about the project, as well as a reality television show of the launch, landing, living, and, presumably, subsequent dying. Even the company's own book, *Mars One: Humanity's Next Great Adventure,* was short on details, with no discussion of the technical concerns of safely flying for months in space, safely landing massive objects on Mars larger than anything NASA had yet landed, and then living in some self-sustaining manner.[18]

Proposed in 2012, Mars One was either an elaborate hoax or an earnest project destined to fail. The company behind it, Mars One Ventures, was merely coordinating an effort to get humans to Mars; the company itself didn't have the aerospace engineering experience and was outsourcing the rockets, spacecraft, landers, habitats, life-support systems—the works. Tens of thousands of people signed up to be the first Mars astronauts through this project, though. The company whittled this number down to a hundred, with the goal of finding twenty-four candidates. Some were married, with children, and were to leave the family behind. (Honey, can you take a hint?) The culling of candidates is all that Mars One had to show for its efforts. Goal one—to launch a communication satellite and supplies to Mars by 2016—hadn't been met by 2020. So, all other goals—rovers by 2018, life-support units by 2020, first crew in 2023—were or would have been missed. Aerospace insiders and experts were largely in agreement that Mars One was theater and would, in the end, never fly any of its components. The company went bankrupt in 2019. I'll say this about Mars One, though: it proved that there is vast public interest in Mars.

Elon Musk's plan for sending a hundred people to Mars in a large spacecraft is only marginally more practical, truth be told. The primary selling point of this plan is that it has Elon Musk's name on it. Unlike Mars One Ventures, Musk and his company SpaceX have proven success in launching rockets. Moreover, SpaceX has a well-articulated plan for developing a Mars transportation infrastructure that includes a reusable rocket now in development to lift 150 tons into orbit, at least 10 tons more than the Saturn V rockets that took humans to the Moon. There is, actually, no current rocket that can propel two humans to Mars, let alone a hundred. The SpaceX plan is to build up to this capability, first with a rocket for LEO and then for Mars. The Mars rocket to carry people would lift 300 tons, including a spacecraft dubbed the Starship configured to have forty cabins and large common areas to get a million people to Mars in batches of one hundred. Do the math, though: that's 10,000 flights. Lotta rockets; lotta fuel; lotta noise. At a rate of a flight per day, it would take more than twenty-seven years to accomplish.

This SpaceX plan is bold but not particularly feasible until several elements are in place. To build a rocket system far larger than the Saturn V, which itself is the largest ever built, would require an enormous investment of tens of billions of dollars, and SpaceX would have to secure that capital. That's not trivial, considering the high risk and low return on investment. Presumably, SpaceX would profit from spin-off technology, namely more efficient smaller rockets to transfer people to LEO or to move cargo around the world in thirty minutes. Next, the rocket needs to be refueled in orbit, a smart idea but untested, nonetheless. The lander needs to be fueled at Mars, and this requires a Martian fuel delivery system on a scale far greater than what's envisioned for the Mars Direct plan—that is, tons of methane and oxygen manufactured robotically on Mars. SpaceX needs to land this monstrosity on Mars, too. Nothing even close to that size has ever been landed on Mars before. It would be more prudent for SpaceX to invest in an Earth–Mars cycler, or two, to transport the masses to Mars, if that's the true goal. Also, a million people is too ambitious and has an air of P. T. Barnum to it.

Because the hardest part of going to Mars is returning from Mars, many big thinkers have called for one-way missions. This includes Buzz Aldrin, the second man to walk on the Moon, and Lawrence Krauss, a theoretical physicist and cosmologist. Krauss discussed the topic in a 2009 op-ed piece in the *New York Times,* in which he suggested sending older people who would live out their days on Mars. Aldrin doesn't consider it a suicide mission but rather one with no immediate *promise* of a return. Should technology improve, we could retrieve the astronauts then. In the meantime, he says, we've made the first steps on Mars. That's befuddling, in my opinion. Why the rush? There's no immediate danger to Earth. Why not send astronauts to Mars when we know how to safely do so? Five hundred years from now, it will make little difference whether the first people arrived to Mars in 2020 or 2080.

In some ways, it's frivolous to speculate about proposed missions to Mars because, since 1952, there have been more than seventy such missions under consideration by governments and scientific societies. And we're not on Mars. That said, we are now closer to landing on Mars than at any other time in history, however much in earnest some of these earlier programs were pursued. Never before have so many governments, companies, and private foundations been so aligned in developing the infrastructure—habitats, suits, greenhouses, vehicles, and more—to live on the Moon or Mars.

NASA is pursuing a long-term, lower-risk, incremental path to Mars. The plan has changed several times since the mid-1990s, being reborn with President Obama in 2009 and tweaked several times since then. So, the NASA plan is difficult to define here and is complicated by the fact that several scenarios are simultaneously under consideration: some involving the Moon in different ways, some going straight to Mars, and some going to Mars with either a four- or six-person crew.* The gist is that there would be a first phase, before deploying astronauts, of sending a descent / ascent

*The NASA roadmap to Mars, "Human Exploration of Mars Design Reference Architecture 5.0," first published in 2009, is 100 pages; the follow-up, published in 2014, is 500 pages.

lander, related surface power systems, possibly separate cargo landers, and a surface habitat. In NASA vernacular, the surface habitat is called the SHAB. As with the ISS, these items might be created in segments on Earth and assembled in LEO. Once tested, off they'd go to Mars. The lander and power systems would land on Mars, while the SHAB would stay in orbit around Mars. NASA has had a string of successes in landing durable robotic devices on Mars and placing satellites around Mars, so all of this is feasible. Once NASA is sure the lander and SHAB are working properly, the agency would send the astronauts in a yet-to-be-designed Mars transfer vehicle in a so-called fast-transit trajectory of about 200 days. The astronauts would rendezvous with the SHAB in orbit and ride that down to Mars. The astronauts would then explore the red planet for forty days, sixty days, or eighteen months, again, yet to be decided. They then would take the descent / ascent lander back to their original Mars transport vehicle, which would have remained in orbit during their stay.[19] Then it's back to Earth for the ticker-tape parade, assuming ticker tape exists circa 2040.

As you might have noticed, the goal relies on several steps that need to be accomplished, and a delay in any one could jeopardize the goal, much like the space shuttle delays doomed Skylab, postponed many missions, and led to ISS cost overruns. Also, this is a one-shot deal to Mars. NASA could repeat it, but the plan itself is not inherently cyclical, aside from leaving some objects behind for possible reuse. The beauty of Zubrin's Mars Direct approach, in my opinion, is that it sets up a system of expanding the Martian base with each two-year landing. Astronauts come and go, but the base gets larger and larger to accommodate a permanent presence of Mars. NASA is considering a return rocket propellant of methane / oxygen like Zubrin, instead of nuclear thermal propulsion, as the plan had been for years. That's one way in which NASA Mars 2040 is looking more and more like Zubrin Mars 2000.

China has similar plans to land humans on Mars while apparently playing catch-up to NASA's accomplishments. This includes a bold plan of simultaneously sending seven orbiters and six rovers by the early 2020s, almost all at once; space stations in LEO and cislunar orbit in the mid-2020s; a sample-return mission from Mars by the end of the 2020s; and multiple missions to the Moon in the 2020s and 2030s to prepare for a human trip to Mars in 2040. Meanwhile, Japan has articulated robotic missions to the moons of Mars by 2024 to support human bases on the red planet later.

Temperature, Pressure, Radiation

Living on Mars, at least for the short term, should not be exceedingly difficult if planned well.

We'll need to set the gravity concern aside, because we just don't know what the long-term effects of 0.38G will be. One thing is clear, though: if astronauts arrive in a spacecraft with a simulated Earth 1G, they will possess superhuman strength on Mars and be capable of lifting, jumping, and throwing three times more than what they could ever muster on Earth. If they arrive after nine months at 0G, they will need about a week to acclimate to 0.38G. Depending on the degree of muscle and bone loss from the journey, they still might feel slightly stronger on Mars.

Temperature should not be a showstopping concern. It will be cold, of course, but still within the range of human comprehension and experience. On Mars, at the equator, it can be as high as 20°C during the day, a pleasant day indeed. At night, with a lack of thick atmosphere to trap heat, the temperatures dip down to –100°C, slightly colder than the Earth's South Pole but manageable.

Fortunately, "night" really is just a night, about twelve hours. A full day on Mars is remarkably similar to Earth's, twenty-four hours and thirty-seven minutes.

The real weather hazard on Mars is a dust storm. The book and movie *The Martian* opens with a dust storm wreaking havoc on the NASA crew and forcing an evacuation. In reality, debris cannot fly around so fast on Mars because there's so little atmosphere to carry it, an inaccuracy that author Andy Weir admits to and, frankly, the only major one in a remarkably accurate sci-fi story.* The danger with dust storms is that they can go global and block out the Sun for weeks or months. Visibility can drop to just a few meters, and dust can coat solar panels on instruments, knocking out power. Dust storms are not uncommon. As bad luck would have it, the Soviet Mars 2 and Mars 3 probes sent down landers at the start of one such large dust storm, in November 1971, greatly compromising the missions and ending them early. NASA's Mariner 9 arrived just one month before but, remaining in orbit, was able to weather out the storm and complete its mission of mapping Mars and conducting atmospheric studies. Nevertheless, a coincidental dust storm upon human touchdown on a frigid, foreign world may very well spell death for the space-weary crew. A long dust storm in 2018 ultimately killed the long-lived *Opportunity* rover, having deprived the vehicle of solar power for months.

Radiation exposure is not insignificant. Mars has no magnetosphere and only a thin atmosphere and thus offers little protection from cosmic and solar radiation. Cosmic rays come from every direction, but the planet itself blocks the cosmic rays hitting the other side of the planet, essentially cutting cosmic radiation exposure in half compared with what astronauts had on their space-

* I'll get to the potatoes later.

craft to Mars. The atmosphere can shield a little more. An experiment riding on the NASA *Curiosity* rover found that the radiation dose rate varied between 180 and 225 μGy / day over the course of 300 Martian days.[20] A "Gy," or gray, is a more health-oriented unit measuring radiation absorption, as cosmic rays are more energetic and thus more penetrating and deadlier than solar radiation, packing more punch. The good news, relatively speaking, is that this is about the same radiation exposure that astronauts encounter on the ISS—not ideal, but not horrible. Humans on the surface of Mars would receive warning of any serious solar flare and could take shelter accordingly. Also, their days might be limited to perhaps eight hours outdoors at the most, whereas the *Curiosity* was exposed 24 / 7 . . . or 24.6167 / 7, to be more precise.

The thin atmosphere on Mars also means that humans would need a pressurized suit to walk about the planet. The average air pressure on the surface of Mars is 6 millibars, and much less on high plateaus and mountains. On Earth, the air pressure at sea level is 1,000 millibars. Even on Mount Everest, the pressure is still a barely livable 340 millibars. But at 6 millibars, your blood would "boil" as liquid turns to a gaseous state within seconds. So, on Mars, regardless of the temperature, you would never, ever be able to walk freely—never be naked outdoors, never feel the real wind in your face—until a radical terraforming effort raises the air pressure to at least 300 millibars. Until that distant day, there's only the ungainly spacesuit to protect you.

Spacesuits, however, might be getting a makeover. They haven't changed much since the Apollo era. Modern ones designed for the space shuttle program and ISS are called Extravehicular Mobility Units, or EMUs, in NASA vernacular. The Russian equivalent is the Orlan spacesuit, *orlan* being Russian for sea eagle. They are clunky, as they are essentially a wearable spaceship, replete with adequate pressure, oxygen, warmth, diaper (spacewalks are long), and a communication system. Astronauts across the years have complained that the spacesuits are difficult to maneuver in and take an hour or more to put on. It's a thirty-step process. Much of that time is spent getting acclimated to the spacesuit "environment," which is 100 percent oxygen at about 350 millibars of pressure.* After the spacewalk, the astronauts need to repeat the thirty steps in reverse.

EMUs won't work on Mars for two primary reasons. First, they are twice as heavy as the Apollo spacesuits; and while that doesn't matter in microgravity, it will matter on Mars. Second, they are far too cumbersome to allow for serious exploration of the red planet in terms of walking, digging, carrying, and such. Every day is a spacewalk day when you're on Mars; that's why humans will be there, after all. EMUs are designed for floating, really, not walking. You don't actually walk or bend your knees much in space. And, let's face it, EMUs lack that certain *je ne sais quoi* of future space exploration.

One promising new approach is based on an old, pre-Apollo design from NASA and the US Air Force called the space activity suit, or SAS, which uses mechanical pressure instead of inflated air pressure to compress the body.† Massachusetts Institute of Technology is developing a SAS that's skintight by design, with thousands of flexible wires that maintain pressure but allow freedom of movement. Led by MIT's Dava Newman, who had held a position at NASA for several years, the engineering team partnered with an Italian designer to ensure that form meets function meets

* You would barely be able to breathe at 350 millibars if the air was only 15 to 20 percent oxygen, as it is on Earth, but 100 percent oxygen ensures comfortable breathing.

† How telling, these acronyms. Would you rather walk around in an EMU or SAS? The SAS seems sassy; the former is for the birds.

style. They got the form part down; the suit looks great. But it's still nearly completely functionless after years of research. Despite the TED Talks and positive press coverage in the mid-2010s, the spacesuit is years away from practical application.

Martian Architecture

The shelter for the first sets of astronauts on Mars would be prefabricated on Earth and sent to Mars either in advance of humans or as part of their spacecraft. Land enough of these habitats, and rather quickly you have a village that rivals scientific bases on Antarctica. Auxiliary shelters could be lightweight, inflatable structures that would be covered with Mars regolith for radiation protection, as is planned for the Moon. NASA has one prototype for Mars called the Ice Home, a giant dome-shaped inflatable that's roomy enough for a four-person crew with such luxuries as a library and growing chamber. The "ice" refers to the several meters of ice and water within the walls and on top of the shelter serving as both insulation and radiation protection.

During humanity's early days on Mars, astronauts would be spending only a year or so on the planet, so long-term radiation protection isn't as crucial for them as it would be for permanent Martian settlers. Long-term habitation will need to be mostly underground or in caves or lava tubes. The surface of Mars is sterile as a result of the radiation. Barring a breakthrough in materials science, there is no practical way to provide an entire community with protection from radiation unless the town is underground or otherwise covered with something substantial. The reason is that protection is closely tied to mass. You wear a lead apron during a dental X-ray because lead is dense and the necessary mass for protection can be held in a thin apron. You could safely wear an equal

An artist's rendering of the Mars Ice Home concept. Home sweet home for a year abroad, the first dwellings on Mars will be Spartan. From the Moon to Mars to Pluto, and beyond, all shelters will be essentially the same in providing warmth, oxygen, and air pressure.

mass of feathers; the apron would just be ridiculously thick, although this would, admittedly, make going to the dentist more fun. On Earth, we have an atmosphere above us blocking most harmful radiation because it has a mass (or pressure) of about 15 pounds per square inch. That's a lot of atoms for solar and cosmic particles to get through. On Mars, the thin atmosphere has a mass (or pressure) of about 0.087 psi. We need much more mass above us, at all times, to attain Earth-like protection from radiation. That mass could come in the form of lead or feathers or dirt or water, whichever is the most practical.

The most practical material for large-scale, long-term habitation would be dirt, or regolith. Lead isn't practical because we'd need to locate, mine, and then smelt the ore; and, anyway, we now know lead is a neurotoxin. Feathers—won't go there. Water is very useful, which is why NASA is considering the Ice Home habitat. A thinner amount is needed for radiation protection compared with regolith. But water on Mars may be too precious to use for this purpose of large-scale and long-term habitats, at least at first. That leaves regolith, which is everywhere. To match Earth's atmosphere's protection, a shelter needs at least five meters, or fifteen feet, of regolith on top of it.[21]

Of course, digging two stories' worth of dirt is not trivial, even on Earth. The first arrivals to Mars would need an excavator for the job. Conceivably, small robotic diggers could do this over the course of a few years. One serious challenge would be contending with water ice, known to be underground on Mars. Any drill or digger would create friction, which produces heat, which would melt the ice. On low-pressure Mars, however, the melted ice would instantly vaporize, not turn to water. If the local atmospheric temperature is below zero, that vapor could then solidify around the tool and freeze it in place. And there would be no one there to bang it free.

The shelter itself, once dug, needs to be (and stay!) pressurized, so one logical material for inside the excavation would be durable, inflatable structures sent from Earth until Mars settlers have the infrastructure to make plastics and mine metals. Shelters also could be built from bricks made with regolith, but they would need to be glazed in some unique way so as not to lose air and pressure. Given these challenges, the first Martian villages might cluster in regions with caves or lava tubes. Here, the hole is already dug for us. The settlers only need to set up their pressurized habitats in these holes. As on the Moon, lava tubes are common on Mars, and some are massive enough to hold entire cities.

You may ask yourself, why go all the way to Mars just to be forced to live underground? Those surface dwellings you see in illustrations and described in science fiction would be suitable only for temporary dwellers. Living on Mars for a lifetime implies living underground until a suitable atmosphere is created or until a breakthrough in physics and engineering allows you to live under transparent yet protective glass. But you can get clever and build into a shaded hillside—for example, the northern face in the northern hemisphere. The hill would block nearly all solar particles and significantly further reduce cosmic radiation exposure. In this scenario, the north side of the structure could have strips of thick, reinforced glass to offer a spectacular view of the Martian environment. This would chase away those claustrophobia blues. Relax by the window; just don't put your bed there. But I say "strips of thick glass" because anything large and thin would be blown apart by the pressure differential. The habitat inside would be around 500 millibars; the world outside is about 5 millibars. So, another common trope of Martian architectural design—large panes of glass—simply wouldn't work with known materials.

It's not too hard to envisage hillside dwellings, as they are common on Earth. A similar shelter scenario would have well-protected buildings—underground or, if aboveground, with only a few windows—all connected by tunnels and tubes, much like the city of Minneapolis, Minnesota, with its vast network of skyways and passages connecting most of the downtown region. Massive indoor malls serve as an instructive example, as well, in which the hundreds of retail stores could be seen as hundreds of spaces for living quarters. Underground buildings could be segregated into residential, commercial, industrial, and educational districts, or any given building could be a community unto itself. As on Earth, we likely would see aggressive urban planning designs to establish Martian villages that will work well for many years but will naturally evolve into new concepts as both unforeseen hazards and efficiencies are realized by the very act of living there.

As Martian technology progresses—that is, when the technologies on Earth can be duplicated on Mars—new architectural possibilities will arise. At noted earlier, water is a better radiation protector than regolith, and it is translucent. It is feasible to make water roofs. A cubic meter of water would provide about the same protection as five cubic meters of regolith. Even a water roof that's fifty centimeters, or twenty inches thick, blocks most of the harmful radiation. So, a double-layer plexiglass roof filled with water could be an attractive option. The weight of the water would help contain the pressure inside. Mars has the materials for making plexiglass (polymethyl methacrylate), and the chemistry is straightforward, although roof maintenance would be rather important given the propensity for ultraviolet radiation to wear down this thermoplastic.

Any shelter would need a rigorous system for air exchange. You can't open a window on Mars. As I noted with submarines, humans inhale oxygen and exhale carbon dioxide, and plants do the reverse. Yet all objects off-gas in some way. Without proper air exchange, the shelter would quickly fill with toxic levels of carbon dioxide, carbon monoxide, and other gases. Between submarines and the ISS, we have demonstrated the technology for regulating air flow, although fresh air can be imported. On Mars, though, there's no escape from bad air.

Breathing Nitrogen and Argon

On the topic of air, consider this seldom-stated fact: we breathe in a lot of nitrogen. The air at sea level on Earth is about 78 percent nitrogen gas (N_2), 20 percent oxygen gas (O_2), and 2 percent trace gases such as argon and carbon dioxide. Only the oxygen gets absorbed into our bloodstream, though. The N_2 goes in and out of our lungs. The two nitrogen atoms are so tightly bound that they rarely react with anything. The fact that nitrogen gas makes up about 80 percent of air means it makes up about 75 percent of the air pressure—more or less. Nitrogen has an atomic mass of 7; oxygen has an atomic mass of 8, slightly heavier. The point here is that air isn't empty.* It's dominated by nitrogen, and inert nitrogen gas is very important for air pressure.

The atmosphere of Mars, thin as it is, is 95.3 percent carbon dioxide, 2.7 percent nitrogen gas, 1.6 percent argon gas, and a whiff of oxygen, water vapor, carbon monoxide, and other gases, according to data from the NASA Viking mission.[22] So, in a pressurized Martian habitat, what will constitute the air we breathe? All we animals need is oxygen. But a 100 percent oxygen environment is highly corrosive and flammable. One spark, and the entire habitat blows up. Ideally, we'd want the habitat air to match Earth's air. The problem is that nitrogen is in short supply on Mars. With some work—that is, energy—we could pull N_2 from the Martian atmosphere. Big planet, small habitat, so there should be enough—at first.

Planetary scientist Christopher McKay has suggested a mix of 50 percent N_2, 30 percent argon, and 20 percent O_2.[23] McKay's

*Now you see that the proverbial glass is neither half empty nor half full but rather completely full, with the other half being air.

reasoning is that this nitrogen-to-argon ratio closely matches the natural Martian atmospheric ratio, in which 2.7 percent is nitrogen gas and 1.6 percent argon gas. So, we could use a machine to suck in the Martian air and remove the carbon dioxide. This would leave us with 58 percent N_2, 34 percent argon, 3 percent O_2, and 2 percent carbon monoxide. We need about 20 percent oxygen to breathe comfortably, so adding oxygen and removing toxic carbon monoxide gives us that 50:30:20 ratio, which I'll call the McKay Cocktail. McKay further calculated the energy needed to process 1,700 m^3 of Martian air to get 1 kg of nitrogen and argon mix: 9.4 kilowatt hours.[24] This is roughly the energy equivalent of drying two or three loads of wash in the dryer, which is attainable with a few solar panels on a sunny day. Humans could live in habitats with 500 millibars of pressure, half that at Earth sea level, reducing the gas budget.*

Nevertheless, if settlements grow from hundreds to millions of people, then nitrogen gas—and argon, for that matter—will be precious commodities.

Send In the Farmers

Nitrogen's not just for air pressure. We need it to grow food. So, when the NASA *Curiosity* rover detected nitric oxide (NO) on Mars in 2015, it was like striking gold. The NO likely came from heated nitrate, NO_3, a biologically accessible form of nitrogen, which is different from N_2. Nitrates can be turned into fertilizer, a necessity for farming. In short, the *Curiosity* discovery means the

*Chris McKay co-organized a highly influential conference in 1981 titled The Case for Mars, which led to published meeting proceedings of the same name in 1984, twelve years before Robert Zubrin's book of the same name.

extremely difficult task of farming on Mars is now slightly easier. Now, where to grow the food?

The bioregenerative life support system (BLSS) hydroponic greenhouse discussed in Chapter 4 would work as well on Mars as it would on the Moon. The challenge is growing 100 percent of a large community's food in a series of BLSS greenhouses. The Moon is close enough to the Earth that bulk food can be flown in; fresh food from the BLSS there is really just a complement, as it is during an Antarctic winter. Moreover, it is my prediction that, because of the Moon's low gravity crippling the chance of raising children there, only a few thousand people will live on the Moon at any given time. That's not that many mouths to feed, and there will likely be no great demand for massive lunar farms.

A goal for Mars from the outset will be to establish self-sufficiency in all its forms. This starts with attempting to grow 100 percent of the food to feed the Martian settlers, regardless of the numbers. One way would be with massive underground greenhouses using artificial light. On Earth, spectacular efficiencies have been realized with indoor stacked hydroponics, or vertical farming. That's layer upon layer of vegetation illuminated by LEDs, with computer-controlled temperature, humidity, feeding, and use of wavelengths best suited for growth or fruiting. No weeds, no bugs. The efficiency can churn out an acre's worth of food, mostly leafy greens, in a space about the size of a shipping container. Mushrooms can grow on wood, stems, and other inedible plant matter, adding protein to the mix and greatly increasing the efficiency in turning all plant matter into edible energy, a key feature of the University of Arizona's "Mushrooms for Mars" project.

This all works extremely well on Earth, but there's one issue few are talking about: Where do you get the lightbulbs? Lightbulbs don't last forever, just a year at best. Manufacturing LEDs on Mars is many decades away; until then, you could ship in food almost as easily as lightbulbs. Until Mars is self-sufficient in LEDs or other lighting elements, it can't be self-sufficient in food production using artificially lit greenhouses.

Mars has lots of land, so space efficiency is not an issue. Mars has adequate sunlight, too, about half of what falls on Earth. Along the equator, which is the likely location for the first bases and settlements, the Sun shines for about twelve hours a day at an intensity of about 600 watts / m^2. That's equivalent to the summer sun on Devon Island at 75°N latitude in Canada. If crops can grow there in a greenhouse—and they can—then they should be able to grow on Mars. But these would be cold-season crops such as cabbages, root vegetables, and winter wheat. The limitation is not the temperature; it's the (lack of) sunlight. Summer crops such as tomatoes and melons need lots of sunlight. You can grow a tomato as far north as Alaska—barely, with the help of a greenhouse for warmth to start them early—but that's because the summer daylight can last for eighteen hours or more. On Mars at the equator, the most you'll get is twelve hours, and that sunlight might be too dim to bring a tomato to fruit with only twelve hours a day of light, regardless of the warmth you can provide.

That's OK. In massive greenhouses on Mars under natural sunlight we could grow various staples to survive, including grains, greens, and grubs to feed chickens. Summer crops such as tomatoes and melons would simply have to rely on artificial lighting. These greenhouses, with or without artificial light, would surely be the most popular area of the Martian base or village, an oasis of green warmth. We could also sustain fish farms. Ultra-efficient systems on Earth employing aquaponics—aquaculture plus hydroponics—form a nearly perfect closed loop: nitrogen-rich fish waste filters through bacteria to become fertilizer that feeds the plants. One caveat here is that, some weeks, the Sun doesn't shine brightly because of dust storms. This is a great, open-ended debate: Are dust storms on Mars frequent and long enough to wipe out sheltered plant growth? If so, could supplemental artificial lighting in the greenhouse get us through some dark patches? A particularly intense global dust storm in 2018 darkened the Martian skies for two months.

Another caveat, and it's another big one, is that the Martian regolith is poisonous. It is filled with perchlorates, chlorine-oxygen (ClO_4^-) -based salts and acids. Plants can't grow in the stuff, and humans will get sick if we ingest it because it compromises the thyroid's ability to regulate hormones. But as the adage goes, when life gives you ClO_4^-, make O_2. A group led by our friend Chris McKay has proposed the solution of using perchlorate-reducing bacteria, or enzymes produced by them, to make the conversion. As proof of concept, the group converted 6 kg of ClO_4^- into one hour's worth of breathable oxygen using enzymes and water.[25] The scientists envision this as an emergency air supply on Mars; just keep a pack of enzymes on hand. Whether this could scale up to make acres and acres of land nontoxic is not known.

Minus perchlorates, crops likely can grow directly in the Martian regolith with fertilizer. A group from Villanova University near Philadelphia, Pennsylvania, tested a variety of crops in simulated regolith. Basil, kale, hops, onions, garlic, lettuce, sweet potatoes, and mint all thrived. Lead investigator Edward Guinan joked the crops might have done even better if the students had remembered to water them regularly.[26] Still, this will be no Garden of Eden. We take much of agriculture for granted on Earth, such as the complex web of life known as soil, a mixture of minerals, dead organic matter, and untold living microscopic and macroscopic organisms. Mars has no soil. So, although plants might grow, their nutritional value remains yet another unknown. Even on Earth, a lack of soil minerals can lead to nutritional deficiencies, such as Keshan disease, once prevalent in parts of China where soils lacked selenium. Visitors to Mars could survive on food grown with the common fertilizer trio of nitrogen, phosphorus, and potassium, or NPK, but settlers would need to be sure they are ingesting the full range of essential vitamins, minerals, and phytonutrients—a range that's not fully understood.

Another missing ingredient is fat. Seed and nut oil could be available, but a lot of space needs to be devoted to this. Sunflowers might suffice. A crude estimate traded among homesteaders is 1.5 kg of seeds for 1 liter of oil, or about a month's supply for a family. A full hectare on Earth could yield 1,800 kg of seeds, or 1,200 liters of oil. That's about a year supply for a hundred families, so this is doable in a larger greenhouse, however challenging. The same applies to rape seed, sesame seed, and peanuts. Tree nuts tend to have a higher oil-to-seed ratio but may be difficult to grow under Martian sunlight.

So, How About Potatoes?

Given the sunlight, soil, and other growing issues, could the character Mark Watney in the book and movie *The Martian* have grown potatoes? Nope. Not in that toxic dirt, and not under those lights. He did everything else right: fertilizer (human feces), water, and a little carbon dioxide. But he would have had to wash the regolith free of perchlorate salts. And those lights, designed for basic illumination, would not have provided enough energy to produce tubers. At best, just a little green growth. Nor would Watney have been able to survive for four years only on potatoes, as was his plan. Potatoes are high in vitamin C, potassium, magnesium, iodine, and some B vitamins, although the nutritional profile of potatoes grown in lifeless regolith and feces has not been established. However, even on Earth, potatoes lack key nutrients. Within a year Watney would have developed a host of symptoms: night blindness from lack of vitamin A, rickets from lack of vitamin D, nerve damage from lack of vitamin E, easy bruising from lack of vitamin K, weak bones from lack of calcium, and a weak heart and deadly Keshan disease from lack of selenium. Also, potatoes have nearly no fat, another essential nutrient.

Watney would have been enormously better off if NASA had packed sweet potatoes instead of standard potatoes. Sweet potatoes are just as easy to grow (under proper lighting); yield more calories per square foot; have edible greens, nearly doubling the nutritional offering of white potatoes; can be eaten raw; and store just as long. I'll go a step further and state that potatoes would not be a top choice on Mars because of their relative lack of nutritional punch and susceptibility to myriad bacterial, viral, and fungal diseases.

I propose that ideal crops to grow in Mars regolith, to complement a hydroponic system, could include cassava, sorghum, cattail, bamboo, and so-called weeds such as dandelion. The reason is the bang for the buck. Cassava, or yuca, grows in poor soil, is one of the most drought-resistant crops known, and has already been made more nutritious with genetic engineering. Sorghum is a grain that can yield a prodigious amount of food in poor conditions on little land. Compared to Watney's potatoes, sorghum has five times the protein, thirty times the fat, and about four times the calories per portion. Dandelion, chicory, lambsquarters, and the like can grow in sidewalk cracks, and every part—root, stem, leaf, flower, and seed—is edible. Similarly, fast-growing bamboo is an ideal building material. As for cattails, no plant produces more edible carbohydrates per acre.[27] Its fluff can be used as stuffing; its fiber can be used for string.

Of course, scientists are thinking of genetically engineered plants and algae that would do well in the Martian environment, even unsheltered, but all of these are untested because no one has access to the true environment, just simulations. We'd be wise to start our farms on Mars with the most vigorous and productive edible plants known on Earth. Kudzu, anyone?

Fortunately, there's water on Mars to grow crops. We now know there is enough water ice, mostly underground, to cover the entire planet in a shallow ocean if it were somehow fully released and melted (assuming an atmosphere to prevent it from vaporizing). In July 2018, the Italian Space Agency announced the discovery of a massive, liquid lake about 20 kilometers (12 miles) wide and 1.5 kilometers (1 mile) below the Martian south pole, which greatly increases the chance of finding life on Mars.[28] A spectacular find, this was the first detection of a massive liquid water source on Mars.*

Send In the Chemists and Engineers

There's currently not much on Mars to build with aside from regolith and the scattering of crashed spacecraft and exhausted landers and rovers. The first visitors will bring stuff: a habitat, a vehicle or two, and a few machines. For the most part, however, we will need to live off the land to make Mars a sustainable operation. And that requires chemists and engineers who can create a civilization literally out of thin air and a hole in the ground.

The need for highly skilled, resourceful, and determined people on Mars is part of what made the Mars One plan so silly. So-called

*For many, this was a double discovery: the presence of liquid water, and the existence of the Italian Space Agency.

ordinary people will be useless on the red planet. Immigrating to Mars today is different from immigrating to the New World in the 1600s in two distinct ways: there are no indigenous people on Mars to teach the new settlers how to survive; and so few people alive today actually know how things work. Most of us in this modern era are capable only of buying and replacing, not making. You may know how to use a computer, but can you make a computer? If your car breaks down, would you know how to fix it with parts you design and fabricate on a 3-D printer?

Countless self-reliance skills have been lost since the last frontier days. How many among us can build a shelter that won't collapse and that doesn't leak, work with metal, turn sand into glass, turn dirt into ceramics, build furniture, prune a fruit tree, preserve food, make vinegar or alcohol, make soap, weave clothing, repair broken machines, understand plumbing and air flow, stop heavy bleeding, splint a broken bone, and so on? The most important people on Mars at first will be those with a deep knowledge of how things work. All of the chemical elements needed for a comfortable, modern life exist on Mars; it's a matter of knowledgeable people manipulating the elements for practical purposes. Robert Zubrin sums it up as a necessity for a broad range of civil, agricultural, chemical, and industrial engineering techniques to turn raw materials into food, fuel, ceramics, glass, plastics, metals, wires, structures, and tools for survival.[29]

High-tech factories are not needed to make these products. Humans have known how to make most of these products for millennia. Only plastic is relatively new, yet plastic manufacturing is a straightforward process of first combining carbon monoxide and hydrogen to create ethylene, C_2H_4, which itself isn't a plastic but is the starting point for making most plastics, both soft and rigid. Ingenuity will come into play, as settlers would need to find alternative materials, such as making concrete with sulfur as a binder instead of water, or making glass without lime (which comes from the skeletal remains of ancient marine organisms). For the most part, though, skilled settlers should be able to create nearly everything they need in Spartan industrial areas within the settlement.

High-tech components could be flown in from Earth and assembled on Mars, since they don't constitute much mass. Consider how a clothes washer is mostly a hunk of thin metal with just a small element of electronics. All the parts of a washer other than the electronics could be manufactured on Mars. This could even apply to space-age technology. Martian settlers would have the ability to build and launch satellites almost entirely from Martian resources, aside from the high-tech electronics. But decades and perhaps centuries would be needed to reach full, modern self-sufficiency on Mars.

Mars as Frontier and Trading Partner

Then again, we don't need to transition Mars to autarky per se but, rather, make Earth and Mars mutually reliant on each other for the betterment of humanity. Mars as a frontier, not simply a scientific outpost, has the potential to spark an awakening of the human spirit the likes of which has not been seen since the Renaissance, which itself closely paralleled the age of exploration. No option other than Mars as a frontier would work, actually. Mars would be too expensive to sustain as something similar to Antarctica, with a few scientists and the occasional tourist.

The analogy often made of Earth to Mars is that of the Old World to the New World, in which the latter became a destination for wealth and freedom that grew increasingly accessible with each passing decade from 1492 onward. In the long term, Mars could occupy a unique position in the Solar System as a fully habitable world between Mother Earth and the far less-hospitable yet economically and scientifically interesting asteroid belt and outer planets. This includes being a supply hub between Earth and the scattered islands of asteroids. To extend the analogy, these more-remote places would be like the West Indies, which provided resources to Europe via the Americas. In the short term, Mars could have valuable resources of direct economic interest to Earth.[30]

Trade would certainly be one-sided at first. Possible Martian exports might include precious metals, such as gold, platinum, rare earths, or rare gems. Another is the fusion fuel deuterium, an isotope of hydrogen, which is eight times more common on Mars than it is on the Earth. I talked about mining the Moon and asteroids; Mars might offer a different array of resources. Also, the concentration and extent of lunar helium-3, another potential fusion fuel, is largely unknown, whereas Martian deuterium levels of 833 ppm hydrogen is far more certain.

But mining is highly speculative. Geologists on Mars would need to find a motherlode, a large and easily accessible reservoir of gold or similar valuable ore. Then, and only if the Outer Space Treaty is worked out to allow for profits, the nation that sponsored the prospecting expedition could sell the rights to a mining operation, which then would invest in machines and workers to mine the resource. This would be largely robotic, but humans would be needed. However, it seems it would be difficult to compete with Earth or lunar mining. Consider gold, at about $2,000 per ounce, or about $60 million per ton. It might cost about $10 million to launch a ton from Mars to Earth, so you'll clear $50 million. But up-front costs to set up a mine could be $10 billion. This means you need to sell 200 tons of gold to break even. The largest mines on Earth have reserves in the 1,000-ton range and sell only ten to twenty tons per year. Competing with the big guys and not driving gold prices down, you'd have to operate for ten to twenty years before recouping your investment. That's a long time to wait on a high-risk investment.

The profitability would depend on the price of getting the materials off Mars. This could be handled in two clever ways. The first would be a rocket hopper to the Martian moon Phobos, a nearly spherical rock (likely an errant asteroid) only 22 kilometers in diameter and 6,000 kilometers from Mars. The Moon, is comparison, is 400,000 kilometers from Earth, more than a hop. A mass driver, or electromagnetic catapult, on low-gravity Phobos could fling cargo toward Earth. The cargo might take a while to arrive at Earth, but shipments might not need to be fast as long as they are regular. The other lift idea is the space elevator, impractical on Earth for safety reasons, even if a strong-enough cable could be spun, but very feasible on Mars. The "areostationary" orbit point for Mars is much closer to its surface than the geostationary point is for Earth, 17,000 km compared with 36,000 km. This, plus the fact that Mars has only 38 percent of the gravity of Earth, means that the tensile strength of the cable can be less, and the total length shorter. So, we wouldn't necessarily need exotic carbon nanotubes but rather a strong material that can be mass-produced with current technology, such as Zylon, M5, or possibly Kevlar. Similarly, skyhooks and space tether systems, described in Chapter 3, would be easier to build on Mars than on Earth. It's a matter of investment. These tools become more practical as more cargo and people need to be placed on or lifted off Mars.

Zubrin mentions one more valuable commodity from Mars, and that's intellectual property. Zubrin's prediction is that the frontier environment on Mars would force settlers to become inventors, creating tools and techniques and finding efficiencies that would be useful on Earth. The distance is the key factor here. Any settlement that's closer to civilization, such as on the Moon or on Antarctica, wouldn't force the settlers to innovate.

Zubrin calls it Yankee ingenuity, this notion of self-reliance and inventiveness in the face of challenges, and it is a key element of both his classic *The Case for Mars* and his 2019 book, *The Case for Space*. I think this is oversell, though, and based on Western bias. We cannot assume that Mars will mirror the conditions of the US frontier that led to Yankee ingenuity, as opposed to the dozens of other frontiers that humans have faced in recent centuries, such as in Australia and Canada. For starters, the Americas were populated with people for millennia who were ingenious in their own ways, simply not patent-minded. The frontier itself doesn't generate patent-minded people. Also, Canada was settled with the same European stock that went to the American colonies, but the cultures of these countries today are quite different. The reason may be that the US frontier was something that could be tamed, which imbued Americans with their tenacity, sense of confidence, and the audacity to coin a phrase like "Yankee ingenuity"; whereas in Canada, it was just too darn cold to bother with frontier conquering. Mars, of course, is more like the far northern Canadian frontier—itself not as daunting the jungles of South America or desert of Australia—than like the US frontier. The seeds of so-called Yankee ingenuity may not grow on the red planet. Thus, one has to wonder what the value of patents would be on Mars to justify the funding and establishment of settlements there.

In summary, making money on Mars is far from a sure thing and, honestly, that could hinder the establishment of settlements on Mars for decades, if not centuries.

Crucial Martian Tech Still Needed

Keeping with the frontier analogy, robots on Mars would serve as both workers and beasts of burden. A key technology that needs perfection is the autonomous vehicle, something that could be programmed to navigate across the Martian terrain. Such a vehicle could possess a pressurized unit for two to four people with supplies—water, air, and food—for a week or more. The people work and sleep in thc unit, free from the necessity of driving. In this way, we could explore vast areas of Mars. Although there are no pedestrians or traffic to contend with, the autonomous vehicle would need to remain vigilant for dry riverbeds, ravines, cliffs, rocks, and other "road" hazards. NASA is developing one such vehicle for local exploration, with two pressurized spacesuits dangling from the back into which astronauts could climb from the cabin to quickly detach and explore.

Vehicles also could run errands, such as driving back and forth to a mine or to a water source, delivering or collecting supplies. Some vehicles could actually create real roads by flattening and sintering the regolith. Over time, a vast network of roads could be created with electronic posts planted at regular intervals to guide other vehicles. Among the most important autonomous vehicles would be diggers that could work around the clock to excavate regolith for shelters. All these vehicles would be an advance over the current generation of rovers because they would be programmed or otherwise controlled locally, not by mission control back on Earth, whose commands, traveling at light speed, are still four to twenty-four minutes away. It took NASA's *Opportunity* Mars rover eleven years to roam 26.2 miles, a marathon's length. As impressive as the rover technology is, these vehicles move only inch by inch upon receiving a set of commands from Earth by controllers worried that their precious rovers will get stuck on the sand or fall over an unseen ravine.

Artificial intelligence helps. The primitive AI system on the *Curiosity* rover enables it to identify rocks and other objects of interest and direct its cameras toward them. The next generation of automated vehicles and machines, on Earth as well as on Mars, will be able to think for themselves, albeit in a limited way. Much like the emerging technology of driverless cars, these machines will be able to sense their environment via sight, odor, or touch; analyze that input and compare it to a preprogrammed knowledge bank; and then act accordingly, whether it is to continue, change course, or stop. Machine learning takes this to a yet-higher level, with the ability of the machine to use statistical techniques to make progressively better movements or decisions, as demonstrated here on Earth by voice recognition software.

This is hardly science fiction. At the Google I/O developer festival in May 2018, Google CEO Sundar Pichai demonstrated phone calls placed by the Google Assistant AI to real people, one for a haircut appointment and the other for a dinner reservation. The first worked flawlessly, despite the complexity of the conversation, as the AI listened to the hair salon worker and chose a time that was conditional on the type of service it wanted. The second call was even more impressive, as it was to a non-native-English speaker who explained that the restaurant didn't accept reservations for parties smaller than four; the AI understood the broken English, asked about typical wait times on a Wednesday, and navigated accordingly.

Augmented reality also will ease the strain of living on Mars. Today, we can wear a device that senses the Earth's magnetic field and buzzes gently when you are facing north. Profoundly deaf individuals can perceive sound through the use of cochlear implants; and we are close to providing some element of sight to the blind through a similar electronic interface. It may be that we will be able to extend sight and sensation past the visible spectrum to include infrared or ultraviolet. On Mars, unrealized technologies may someday soothe the brain with the sights, smells, and sounds of oceans, forest, other precious elements of Earth, or enable settlers to interact more closely with friends and relatives back home.

As for 3-D printing, printers could be stationary, like those on Earth, printing out tools and other gadgets from an ink based on regolith or discarded materials from the journey to Mars, such as parachutes, rubber, or plastic. As we saw planned for the Moon, the robotic printers could arrive before humans to sculpt a shelter or ribbon of road. A group led by Ramille Shah at Northwestern University in Chicago has developed a method of creating 3-D printer inks from either lunar or Martian regolith by combining these substances with simple solvents and a biopolymer, resulting in a flexible but tough material similar to rubber, 90 percent regolith by weight.[31] The group made interlocking bricks with the material. A longer-term goal will be to make the process entirely robotic, so that a rover could take a scoop of regolith, make the ink, and load the printer. During our early days on Mars, the labor market will be tight, to say the least, and working outside comes at the price of radiation exposure and time needed to suit up properly. Anything that can be done by a robot—driving, carrying, digging, stacking—needs to be done by a robot.

All this tech is heavy and delicate and needs to land on the planet, calling attention to one more bit of tech that needs to be perfected: the technology of landing heavy technology. We actually don't have a great track record for landing stuff on Mars, although it's been getting better. Mission Control explodes into cheers when they have a success, with grown men kissing and hugging each other, for good reason: more than half the missions to Mars have failed. Many have crashed upon landing. The USSR / Russia has only had two successes in twenty attempts. In the 1990s, four out of six NASA missions to Mars failed. Engineers call it the Mars Defense System. However, since 2000, there have been "only" three failures in twelve missions—although two of the three failures were lander losses. It's safe to say that it's difficult to land objects on Mars. The planet's relatively high gravity and thin atmosphere conspire to complicate aerobraking. Nothing heavier than a ton has ever been landed successfully on Mars.* Those autonomous vehicles? Some weigh several tons.

One entry vehicle under development by NASA is the Low-Density Supersonic Decelerator (LDSD), which looks a bit like a flying saucer. The LDSD would deploy a massive, inflatable Kevlar tube to create atmospheric drag to slow the craft from supersonic speeds, in the range of 6 km / sec, to only 0.5 km / sec in just a few minutes, slow enough for extremely large parachutes to take over in the lower atmosphere, thin as it is. Even this won't make the descent slow enough. Engines would be needed for retropropulsion, firing toward the direction of the ground to slow the craft even more so that it can touch down softly and upright. The NASA Mars Science Laboratory (MSL), carrying the 899 kg *Curiosity* rover, entered Mars almost in this fashion. The MSL had a 4.5-meter-diameter heat shield for initial braking in lieu of the LDSD. The shield reached a peak temperature of 2,090°C, hot enough to melt iron.[32] The shield, the largest ever flown, wouldn't have been able to protect anything much heavier. The LDSD is being tested for loads upward of three tons.

*The USSR's Mars 3 was 1,200 kg and is said to have achieved a soft landing, but the probe failed fourteen seconds later.

Minus a reliable entry vehicle, settling Mars will be postponed indefinitely.

Martian settlers would need to perfect closed-loop systems, too. The ISS and nuclear submarines regularly get a fresh influx of supplies; and the reality is, no humans have perfected a closed-loop system in which water, oxygen, and carbon dioxide are fully conserved. This attempt failed famously for Biosphere 2. The approach on Mars would be to start modestly and have an emergency backup of the essentials: oxygen and water. Biosphere 2 was an ambitious experiment, with five different Earth biomes. Habitats on Mars need not be complexified with jungles, wetlands, and coastal environments. They should, however, be compartmental, so that any collapse of one area—such as unexpected bacterial growth consuming oxygen, as we saw in Biosphere 2—can be quickly isolated and remedied. An utterly ludicrous idea would be any habitat concept taking a prize in contests sponsored by Mars City Design, a collaborative platform for planning Mars architecture, which they call Marschitecture. All razzle-dazzle, the designs attempt to have us fly before we can crawl.[33] Designs for grand architecture on Mars will evolve on Mars, not on Earth. Cities will grow organically from modest beginnings, guided by practicality, not the imagination of people who have never visited the area. So silly. They will simply never come to fruition.

Who Goes?

In the beginning, the travelers to Mars will be people handpicked by government space agencies. Should settlements become feasible, likely not until later this century, the way things are progressing now, then immigrants will be those who either can afford the trip easily or who feel compelled to seek a new life, such as the very early American colonists and other settlers who cashed in most of their holdings on the gamble called the Americas. When you have little to lose and all to gain, the decision might be easy.

Let's explore the scenarios. Several nations have their eyes on Mars. Russia would like to have a presence there, but they remain haunted by so many failures. Russia has never fully successfully landing anything on Mars or its moon. Even in the modern era, its 2011 Phobos-Grunt / Yinghuo-1 mission with China never made it out of low-earth orbit. Its 2016 ExoMars Orbiter / Schiaparelli EDL Demo Lander mission with ESA was a partial success: the orbiter made it, but the lander crashed. Russia has loose plans for reaching Mars with humans by 2045, but the Russian Space Agency has no announced plans for any robotic mission between now and then. China, as mentioned, has a multitude of missions planned. A manned mission could happen by 2035. And it likely will be "manned," without women, because China (and Russia) seem averse, for cultural reasons, to sending a mixed-gender crew on such a long and intimate mission. Although the United States is far more experienced in reaching Mars, China is on a trajectory to beat the United States in sending humans there, and it may be that US commitment to the effort won't come into focus until the "threat" of China getting there first is more fully realized.

NASA is conducting research on the best crew to send to Mars. The group would be four to six people, likely comprising both men and women, particularly if more than four go. A four-person crew very well could be all male. Sending an all-female group has practical merit: women consume fewer calories and thus need less food (i.e., less mass, fuel, money); and women under age thirty are better protected from radiation-induced endothelial and vascular damage than are men.[34] The uncertainty of the sexual dynamics of a mixed-gender crew and its effect on the mission is driving a scenario of a four-man mission or a four-man / two-woman mission and not a balanced gender ratio—a LUCA-generating topic, to be sure.

NASA doesn't fully reveal its criteria for choosing astronauts, mainly because the space agency doesn't want someone gaming the system to be chosen. It is safe to say, though, that NASA will use a separate set of criteria for selecting astronauts for a Mars mission than it would for an ISS or Moon mission. During the Mercury, Gemini, and Apollo years, it was all about type-A personalities: driven, athletic, competitive. Once access to space became more predictable, the focus changed from military pilots with technical expertise and leadership skills to scientists and engineers with a diversity of skills, as exemplified on the ISS. The journey to Mars, however, will be a new challenge, requiring a unique personality profile.

First, let's look at the mission profile. A mission to Mars will last two to three years and will be defined by cramped quarters, monotony, a sense of isolation, and physical challenges. As such, NASA is considering candidates who can meet these challenges yet also complement each other in terms of personality and skills.⋆ Personality traits will include some combination of the traditional five-factor, or Big Five, model of personality traits used for submarine mission selection: degree of neuroticism (inversely, emotional stability), extraversion, openness, agreeableness, and conscientiousness.[35] On a long mission to Mars, some degree of introversion might be appropriate, too—that is, someone who thrives being left alone, or at least tolerates it, such as those who can perform

⋆With a mission planned for the 2040s, most candidates will likely be children at the time of this book's publication.

science during a winter in Antarctica and stay sane. Related to the Big Five traits are resilience, curiosity, and creativity. Lightheartedness may be an important trait, as well. Tristan Bassingthwaighte, a member of HI-SEAS IV project in Hawai'i, told me he may have been chosen for this Mars analog because of his personality traits: funny and pleasant. Yet knowing what will work best is a bit of a guessing game, with no way to test in advance and ultimately no one to validate the choices, unless dozens of missions to Mars are flown with different sets of personality traits.

As for skill sets, this one is clearer. Each crew member needs to be able-bodied, well-educated, and intelligent, with strong expertise in one mission element yet knowledgeable about every aspect of the mission both in-flight and on Mars. Assuming a stay on Mars of at least several months, the first crew must have at least one person expert in each area: flight, biology, geology, chemical engineering, and mechanical engineering. This core set of knowledge will help ensure the crew can get there safely (the pilot), set up shop (the engineers), and study Mars (the scientists). There's often an overlap between pilots and engineers. Beyond this, two of them need to be also trained as medics, in case one medic is injured. Redundancy is prudent for all other secondary skills, such as food preparation, gardening, and computer or electronic maintenance.

Remember, NASA, ESA, and other space agencies are not (yet) thinking about a long-term settlement. The first few missions would be entirely scientific, setting up a base in so-called exploration zones with no intent of joining bases to create a village. Proposed zones are scattered across Mars. Some bases might be abandoned; others might be visited a few times by different crews. NASA sees its role on Mars as exploratory, a little like the explorers of the sixteenth and seventeenth centuries. Settlements or colonies would come later, directed or inspired by information and lessons learned by government space agencies but otherwise funded and executed separately. If it turns out that there's little of value on Mars, after all, or if some unexpected hazard would make a permanent settlement too difficult—such as a toxin or the low-gravity problem—we might be forced to call off dreams of a second Earth—for this century, anyway.

Let's assume that the first generation of human explorers demonstrates that colonization is feasible.

Who *really* would go? Again, we can take our clues from history. In the Americas, some of the earliest pioneers—before the motivation of escaping religious persecution—were men looking for riches. This included prospectors and trappers who penetrated deep into the wilderness. If mining became profitable, surely both men and women would sign up for extended stays on Mars, not necessarily to spend a lifetime there but save up a lot of money for ten years or so. Some might stay; most would return to Earth eventually on a trade ship.

It's hard to imagine religious persecution driving immigration to Mars, not because religion is no longer relevant but rather because those who are most oppressed—the Rohingya people in Myanmar, for example—tend to be the poorest, unable to afford the journey. And they likely could escape persecution by finding a different country to live in, not a new planet. Nevertheless, ideology could be a driver to Mars. Many people, no doubt millions, are so frustrated with politics and other matters on Earth that they'd opt for total freedom from government on Mars. Many would be enticed by the frontier and the hope that they could secure land and wealth for their prodigy merely by being first to the new world. After all, those families who have been in the United States the longest (excluding Native Americans) tend to be the wealthiest.

I think it will come down to money. Once it becomes affordable to travel to Mars, people will go. Physicist Freeman Dyson has thought about this topic, too. He compares space migration to the other great migrations. He estimated that the *Mayflower* voyage cost the average family about 7.5 years of wages. The Mormon trek to Utah cost about 2.5 years of wages.[36] Currently, a trip to Mars would cost more than a billion dollars, or 10,000 years of wages. No one can afford it. But a trip for a million dollars, or about ten years of wages, might be affordable. Some people buy houses for about ten years of wages, after all. Could we get the price that low? Dyson also calculated in the same essay, titled "Pilgrims, Saints and Spacemen," that 128 years passed between the voyages of Columbus and that of the *Mayflower;* and during that time, nations of Europe were building ships and establishing the commercial infrastructure that would enable the journey of the famous *Mayflower* from Plymouth, England, to Cape Cod, Massachusetts, with 102 passengers. The year 2085 would be 128 years after Sputnik, and I think it is conceivable that travel to Mars could be affordable by that point.

I should note that, of those 102 passengers on the *Mayflower,* only fifty-three survived the first winter. Years before that, at Jamestown, 440 of the 500 colonists died in the winter of 1609–1610, a period known as Starving Time. Mars won't be a picnic for the first settlers, either. There will be deaths.

Who Owns Mars?

There's one more major hurdle to overcome before settling on Mars, and that's the Outer Space Treaty. As I mentioned in Chapter 4, the Outer Space Treaty clearly states that Mars cannot be "subject to national appropriation by claim of sovereignty, by means of use or occupation, or by any other means."

If one chooses to place settlements on Mars, there's little way around this treaty, other than to ignore it. If Elon Musk, for example, follows through to send one million people to Mars, he'd have to set up his company in one of the dozen or so countries that have not signed or ratified this treaty, such as Liechtenstein. The settlers would need to renounce their citizenship—unless they are from Liechtenstein. Ignoring or withdrawing from treaties, of course, is not without precedent. In 2002, the United States pulled out of the Anti-Ballistic Missile Treaty of 1972 so that it could create a missile defense system. More telling, the United States has pulled out of the majority of treaties made with Native American tribes primarily for one reason: money: if not gold, then silver, oil, or other lucrative resources discovered on lands assigned to Native Americans. If highly profitable resources are found on the Moon or Mars, we can assume the Outer Space Treaty will be abandoned. The Outer Space Treaty is also threatened by the increasing calls for militarizing space, as is the apparent intent of both China and the United States, the latter case referring to the proposed Space Force as a sixth arm of the US Department of Defense.

I'm not pessimistic by nature, but I don't see how Mars could be anything but a land grab driven by homesteading rules if scientific explorations demonstrate that permanent settlements are economically sustainable. I predict that, in lieu of a suitable treaty, the first entities to Mars will own Mars. Larger, spacefaring nations might agree to carve out equal, vast areas to administer; or, to prevent any one nation from claiming too much, the United Nations could establish world heritage zones to limit the amount of territory that's open for grabs. Such a treaty could be placed into effect for, say, fifty years and revisited as Martian settlements grow. In short, however effective the Outer Space Treaty was in preventing the nuclear arms race from spiraling into space, it will be abandoned or replaced because it leaves little room for the practical utilization and commercialization of space now that space is more approachable. A carefully revised treaty, allowing for land appropriation, may actually stimulate exploration and colonization.

What If Life Is Found on Mars?

Finding life on Mars changes everything. This would be one of the most profound discoveries in the history of humanity. If life has ever existed on Mars, the implication would be that life must exist throughout the galaxy and universe. Think Isaac Asimov's zero-one-infinity rule: two is an impossible number. Either there's no planet with life, one planet with life, or an infinite number of planets with life—but not two or three planets. The discovery would lead to a raft of more questions: When and how did life arise on Mars? Did life as we know it originate on Mars and hop over to Earth on a meteorite? and, Is there life elsewhere in our Solar System? I think finding life on Mars, even fossilized, would open a new era of space exploration to the moons of Jupiter and Saturn, which also could harbor life.

As for the fate of Martian settlements, the question is even more profound: Now what? Would the establishment of settlements be ethical or even safe for humans in the presence of alien life? My opinion is yes to both points, but of course there are plenty of respectable counterarguments surrounding this topic.

From an ethical point of view, I don't see why our freedom to expand should be limited to Earth. Humans left the African continent about 60,000 years ago with no sense of natural law prohibiting their migration to nearly every other continent and island on Earth. I see Mars as a natural extension of human migration, regardless of what kind of life is there. Some argue that we need to study such life before we inadvertently destroy it. However, if life still exists on Mars, is it living below the surface. Humans on the surface and in isolated subsurface dwellings would hardly come into contact with that life. In fact, ancient microorganisms live far below the surface on Earth; the surface, with its oxygen and light, is toxic to them. We don't touch them, and they don't touch us. On Mars it would be the same unless we encounter a world of underground gnomes, which I think is highly unlikely.

As for the safety, I think biological infection is also highly unlikely. Spurred by sci-fi novels such as *The Andromeda Strain,* some folks are convinced that an alien microorganism from Mars could wipe out human life on Earth. But that's not how biology works. Deadly organisms, such as those causing smallpox or Ebola, coevolved with humans and other living creatures for hundreds of millions of years. Viruses, in particular, multiply by hijacking host DNA. Some viruses, known as phages, attack only bacteria; others affect plants; still others, animals. As deadly as viruses are, they evolve to infect a limited range of hosts. Many people have died from starvation as a result of potato blight, but no human has ever been infected by *Phytophthora infestans,* the oomycete that causes potato blight.

For any microorganism on Mars to be infectious to humans, it would need to find it advantageous to derive nutrients from and reproduce in the human body at 36.1°C instead of in the nearly frozen, underground world in which it has survived, presumably for billions of years. This would be a remarkable coincidence and, really, a bit egocentric to think that an alien microorganism would find humans, among all life forms, to be most desirable. Any threats to human health on Mars would be inorganic, perhaps rashes or lung damage from the Martian dust because of the harsh chemicals it contains. However deadly, this would not be contagious.

I'll close with this: the possibility of finding life on Mars, past or present, is high. Mars was warmer and wetter than it is now for more than a billion years, during the same period in which life arose on Earth. Mars had a protective atmosphere, vast oceans, and a water cycle. Indeed, the red planet had nearly the exact same conditions as Earth had to harbor life. And today, it seems, there's at least one underground lake on Mars with liquid water. Life could have receded from the surface into underground havens, as some life on Earth did 2.7 billion years ago when the air became too toxic with that stuff called oxygen. The way astrophysicist Neil deGrasse Tyson sees it, there's no reason to think that life on Earth is unique. Life as we know it is composed of the most common elements in the universe: hydrogen, carbon, nitrogen, and oxygen. Life on Earth would be unique if it were based on an isotope of bismuth, Tyson likes to say. The fact that life instead contains the most abundant elements released in star explosions suggests that it could be commonplace. If we don't settle Mars solely because life exists there, then we'd be fated to limiting our expansion in the universe only to moons and planets too inhospitable to accommodate life.

Days, Nights, and Holidays on Mars

Welcome to Mars. The White Zone is for loading and unloading only. Please do not leave your spacecraft in the White Zone unattended.

We are decades away from setting foot on Mars, let alone establishing a permanent settlement. Indeed, this chapter rests on three assumptions, which I do think are sound but nevertheless not guaranteed to happen quickly. Assumption one: sometime between 2030 and 2050, humans will reach Mars. Likely this expedition will be US-led or China-led, although it is not inconceivable for it to be led by a private endeavor. Assumption two: the 0.38G is enough gravity in which to reproduce and raise healthy offspring. Assumption three: millions of people will *want* to go to Mars; and among them, thousands will succeed in getting there, a few decades after our initial footsteps.

I believe there will be more interest in settling Mars than in settling Antarctica. The reasons are twofold. First, I cannot deny Mars migration will be driven partially by naïveté. Many of those who want to live on Mars don't truly understand how challenging it is. They will travel to Mars for the novelty more than the practicality. They also will be propelled by the notion of a grand human experiment, never to be eclipsed until, perhaps, we set off for other stars in generational spacecraft. The first generation of settlers would be immortalized. Maybe these are the wrong reasons for going to Mars, but they are reasons nonetheless. Second, a subset of immigrants will set off for Mars for the pursuit of freedom, an opportunity to truly escape Earth governments and to establish their own sense of ideal governance. Of course, there's irony in the fact that the price of freedom will be containment in shelters on Mars.

But let's explore day-to-day life on Mars for the scientist, the tourist, the transient worker, and the permanent settler. One element that makes Mars so inviting is its similarity to Earth in terms of time and seasons. A sidereal day on Mars, the time it takes to revolve a full turn on its axis as measured relative to a fixed star, is twenty-four hours, thirty-seven minutes, and 22.663 seconds. That's coincidentally close to the Earth's sidereal day of twenty-three hours, fifty-six minutes, and 4.096 seconds. So, our biological clocks might adapt to the Martian day–night cycle.* How we handle the slight time difference will be interesting. We could create a Martian second, slightly longer than an Earth second, and keep a twenty-four-hour clock. Seconds, minutes, and hours each would be about 2.7 percent longer. Several NASA missions worked on this time; and some members of the NASA Jet Propulsion Lab wore Martian wristwatches created by master watchmaker Garo Anserlian to keep to a proper schedule for operating NASA rovers.

Another concept would be to start new, get rid of the twenty-four-hour clock, which is built on ancient number systems that were duodecimal (base 12) and sexagesimal (base 60). We could switch to metric and create a day with ten Martian hours. That's ten seconds to the minute, and ten minutes to the hour. The day would have 1,000 Martian seconds, a second that we know would be between a Martian microsecond and millisecond. Or have a single unit, the day, with components called the deciday, centiday, milliday, microday, and nanoday. A new metric unit between milli (10^{-3}) and micro (10^{-6}) at 10^{-5} (let's call it "seco") would mean that one secoday is about 0.8 Earth seconds. This may seem awkward at first, but to the second generation on Mars it would be natural and ultimately make calculations easier.†

*Michael Young, co-recipient of the 2017 Nobel Prize in Physiology or Medicine for "discoveries of molecular mechanisms controlling the circadian rhythm," told me in an interview that the forty-one-minute daily difference actually might be hard to overcome, as visitors grow out of sync with Earth by about five hours each week. I found that hard to believe, but I haven't won a Nobel Prize.

†Similarly, we could switch to base 12 counting on Mars and base the metric system on that. The five most elementary fractions (½, ⅓, ⅔, ¼, ¾) would simply have duodecimal renderings: 0.6, 0.4, 0.8, 0.3, and 0.9, respectively. No more 0.3333333 and so on for ⅓.

One Martian day is called a sol. One Martian year, or a full journey around the Sun, is 668.60 sols or 686.98 Earth days. That's 1.9 times longer than an Earth year, again coincidentally and conveniently close to two times, or double. So, you can say that a day on Mars is like a day on Earth, and the year is twice as long. On no other planet or moon does this convenience hold. Moreover, Mars tilts on its axis at 25 degrees, very close to the Earth's 23.5-degree tilt. This means Mars has four seasons, although not evenly spaced as they are on Earth because Mars's orbit is more elliptical. From a northern hemisphere perspective, spring is about 7 months long, summer is 6 months, fall is 5.3 months, and winter is about 4 months.

Scientists would have a field day on Mars, and there's really no end to the types of field science that can be conducted there. High on the list would be searching for life. Everything else may sound mundane in comparison but is crucial, nonetheless, including (1) understanding what happened on Mars two billion years ago when it went from warmer and wetter to cold and dry, and (2) characterizing the geology of Mars. That's basic biology, climatology, chemistry, and geology—or "areology." An early goal will be establishing the infrastructure to support yet more exploration, as well as surveying the terrain for resource potentials. Much of any given day in the early years will be spent on building and maintenance.

Tourism could follow if commercial enterprises learn how to lower the cost and quicken the travel time. By the time the first humans get to Mars, circa 2040, we most certainly will have established tourism in low-Earth orbit and quite likely in lunar orbit and on the Moon. More so than the price tag, though, the obstacle for the Martian tourism trade to overcome will be the distance to the red planet. The Moon we can get to in less than a week. A wealthy thrill seeker need only set aside a few weeks for a brilliant lunar vacation. Barring a propulsion breakthrough that would shorten the Mars trip to a few weeks—feasible but unlikely this century—visitors to Mars would need to set aside two years of their life for this experience. The cost, while certainly not affordable for 99.99 percent of the population, could be in the few tens of millions of dollars and attractive to the super-rich. If you do the math with me, the fare for ten of these travelers on a twelve-person trip to Mars (includes two crew members), each spending $100 million for the thrill, could add up to $1 billion, a reasonable cost estimate for a journey occurring after infrastructure is in place on Mars.

Just being on Mars would be 99 percent of the tourist experience, and my acting as a travel agent here, telling you the best sites to visit, is a bit premature. Mars is an entire planet, after all, about half the size of Earth; and traveling from spot to spot on Mars in a pre-infrastructure era will be essentially impossible. It is possible to fly on Mars in dirigibles and airplane, but such comforts will take some time to establish. Until then, when visiting Mars, you would be confined to a roughly 100 km radius that can be reached in a pressurized tour bus. One must-see site would be Valles Marineris, among the largest canyons in the Solar System and arguably the most dramatic. Valles Marineris is 4,000 km long and up to 200 km wide and 7 km deep. Earth's Grand Canyon is "only" a tenth of this size. No camera has captured high-resolution images of Valles Marineris; we can only imagine the splendor. If you camp out on the western edge, you should have a view of Tharsis Montes, about 1,000 km to the west. This trio of evenly spaced and massive shield volcanoes was formed from lava. The tallest is Ascraeus Mons, with a summit elevation of more than 18 km. Mount Everest, the highest peak on Earth, is about 9 km.

Beyond Tharsis Montes by another 1,000 km, and perhaps visible from Valles Marineris, is Olympus Mons, the tallest planetary mountain in the Solar System at 25 km in height. Noctis Labyrinthus, a region nearly indistinguishable from the Badlands of South Dakota, is a few hundred kilometers from the western edge of Valles Marineris. And all of this is near the equator, where the weather is the warmest. So, for all these reasons, my pick for a tourist destination would be western Valles Marineris. You'll likely get a view of the stunning polar caps when you fly home.

A transient workforce may accompany tourism. I have already discussed and generally minimized the role of remote mining on Mars in view of better mining opportunities on the Moon and on near-earth asteroids. The distance from Earth (the market) and the need for heavy exports to escape the Martian gravity will conspire to make Martian mining an unattractive investment option. And it's all about investment: one must convince investors to put up money for a promise of return, and this would be a tough sell. I daresay there might even be more environmental concerns about polluting Mars than about ripping apart asteroids or the Moon, which are seen more as barren rocks.

There could be business opportunities on Mars early on, though, akin to what's happening on Antarctica. Governments run bases on Antarctica, but much of the work is contracted out to companies such as Raytheon Polar Services, Gana-A'Yoo Service Corporation, GHG Corporation, PAE, and dozens of others you've likely never heard of. These companies employ hundreds of workers who live and work on-site. Should multinational science bases come to Mars, then it follows that support crews will be needed. Pure speculation here, but workers might sign up for a ten-year stint on Mars, as opposed to a year or two on Antarctica, because of the cost of transit. One cannot come and go so easily to Mars.

Antarctica is not profitable for any nation, aside from the intangible value of knowledge gained and the potential long-term return on investment. The United States has the largest presence there, and it spends about $500 million per year to build and maintain those scientific bases through the US National Science Foundation Office of Polar Programs.[37] That's about 7 percent of the $7.5 billion NSF budget. Science is nice; it surely pays dividends far down the road as it applies to technological advancements. But the real reason the United States and thirty-nine other countries have a presence on Antarctica is geopolitical. They have to be there or else get shut out of any resources that might be extracted from the continent. There's no reason Mars wouldn't be the same. Either become a player or watch the game. The question is, how much investment can any given nation tolerate? The ISS cost the United States $100 billion to construct and continues to cost $4 billion per year to maintain, a real boondoggle, as explained in Chapter 3. Yet this might imply that the United States could tolerate a similar investment on Mars, at that price. That tolerance would be higher, should China have a base or two on Mars.

If there are scientific bases, of course there will be scientists. These would be scientist-astronauts at first but, as travel costs and risks decrease, the workforce will transition to scientists supported by a small crew of regulars. We will see this on the Moon, too, but the "regulars" on Mars would be staying longer. The pay won't need to be high. The first set of workers would go for the thrill; and with little to spend their money on, they'd bank most of their paycheck—again, as is the case on Antarctica. These transient workers would have the arduous, day-to-day task of maintaining the bases and expanding the infrastructure.

Many decades would need to pass between the first humans working on Mars and the introduction of settlers who have the full intention of migrating to Mars and raising children there. There would be the necessary safety checks: primarily, can humans live, grow, and reproduce on Mars? And then there's the price issue: primarily, can we at least lower the cost per family to *Mayflower* cost? Settlements are off until both these answer are yes (unless Mars is settled as some kind of kid-free retirement community, with residents who are not worried by long-term radiation risks and who enjoy the weaker gravity). As for these first permanent settlers, they mostly would be engaged in the endeavor of homesteading. Homesteaders yearn for self-sufficiency and the opportunity to transform challenging terrain into something beautiful and productive. The settlers would spend their days building and performing maintenance, like others on the planet, but also farming. They need to secure not just water, food, shelter, and energy but also oxygen and air pressure, things homesteaders on Earth take for granted. Everything must be carried out with the utmost efficiency because resources are so precious. In time, those resources would form the basis of the Martian trade economy—commodities such as water, oxygen, and nitrogen, along with standard homesteading fare such as food, clothes, detergent, and tools.

Local Sights

Novel experiences on Mars would include gazing up to see two moons, Phobos and Deimos. Phobos, the closer and larger of the two, whips around the planet in seven hours and thirty-nine minutes, so you would see it passing by somewhat quickly three times a day at an apparent size of about a third as large as our Moon. Deimos, a tiny thing only thirteen kilometers wide, goes in and out of view about every sixty-six hours and would look a little larger than Venus does from Earth. Scattered sunlight through the veil of dust in the air paints the daytime sky on Mars a pinkish red and sometimes butterscotch; sunrise and sunset would offer a cool-blue-colored Sun. The Sun would look about 0.7 times smaller than you are used to, and 40 percent as bright.

If you're into astrology—and still think you can handle Mars—note that you'd have to contend with a slight shift in the constellations. The Sun spends six days in the constellation Cetus in the middle of Pisces, complicating romances, no doubt. The winter solstice is in (gasp!) Virgo. The brightest planet would be Jupiter; Venus is visible but dim. Earth would appear as a bright star, although not as bright as we see Venus from Earth. With a small telescope you could see Earth's Moon. The north stars (plural) would be Gamma Cygni and Alpha Cygni, not Polaris. The night sky would otherwise closely resemble what we see from Earth, with only small seasonal changes, perhaps enough to spark the imagination in redefining constellation patterns. With no light pollution and a thin atmosphere, Mars would offer brilliant astronomical seeing, but the dust likely dims all objects by at least an optical magnitude.

The terrain on Mars is diverse and populated by extensive sand dunes, canyons, dry riverbeds, dormant volcanoes, mountains, valleys, and plateaus. Should you go hiking, you would have an extra bounce in each step, as you would weigh less than half of what you do on Earth. Your gait likely would be different once you learn how to move efficiently. In the beginning, though, you'd likely be hitting your head on ceilings quite often. Will these exotic experiences of a life on Mars be appreciated by the second generation, who know not Earth? Perhaps so, but in a different way. Mars has an inherent beauty. But we shouldn't kid ourselves: much of the lure is the novelty, a charm that some will come to regret and that future generations will interpret as normal.

Earth seen from Mars. That dot in the sky represents all of humanity, all of our memories, all of our dreams past and present. Settlers on Mars will need to contend with a profound level of isolation. The photograph was captured by the *Curiosity* Mars rover.

The New Martian Body

No one knows for sure what living on Mars would do to our bodies, but there would surely be a host of changes. In the same way that identical twins raised in different environments would grow in different ways and soon begin to look less identical, humans on Mars will undergo changes that will alter their appearance compared to humans on Earth, perhaps drastically. Radiation, temperature, light exposure and other factors will cause some genes to be expressed and others to be suppressed. These changes would be most dramatic in children raised on Mars but would apply to adults, as well.

The first visible change might be in the eyes and skull. In a world of artificial light and sunlight dimmer than we experience on Earth, the human eye would receive a different set of sensory input instructing it to grow. Exposure to dimmer light at an early age might cause the eyes to become bigger to collect more light. The whole skull might change to accommodate a larger eye socket. Adults who migrate to Mars with fully developed eyes also might develop enlarged pupils. The human face is what we know best, so it is in the face that we will most readily perceive changes. For example, you might be able to notice subtle changes in the faces of recent immigrants compared with their children, such as an Asian American compared with an Asian, beyond wardrobe effects. Speaking different words and chewing different foods will alter the shape of the face ever so slightly. If the Asian American returns to Asia and has a child, that child—speaking Asian words, eating Asian food—would assume the original look of an Asian.

Skin tone would change on Mars, too, but scientists are unsure in which direction, lighter or darker. The increased radiation could stimulate the production of eumelanin, leading to darker skin, although this has not happened to fair-skinned Europeans who migrated to sunny Australia.[38] They just get skin cancer. More likely, we will get paler, like anyone who is housebound and not exposed to natural sunlight. Martian dwellers may desire to eat foods with carotenoids, such as carrots and sweet potatoes, as these pigments accumulate in the blood stream and skin and offer some protection against UV damage. In this case, Martian dwellers will develop an orange hue. Another great unknown is the fate of the human microbiome on Mars. The microbiome—skin, gut, and other bodily microbes living in harmony with us—control digestion, nutrient absorption, and the immune system in ways we are only beginning to understand. With limited exposure to microbes on sterile Mars, the human microbiome will be altered in unpredictable ways. This could not only leave humans more vulnerable to infections but also drastically alter our appearance, making us thinner or fatter, for example. Frankly, there's good reason to believe that humans may never thrive on Mars, given the immensely important yet mysterious interaction between us and the bacterial world around us.

In a few generations, evolution and the nascent indication of speciation will begin to kick in. This could be driven predominantly by the founder effect. The genes of the first batch of immigrants will be the ones that most come to dominate Mars, unless there's an immediate and constant flow of migration. With high mortality probable, "survival of the fittest" may become apparent as those with genes better adapted for Mars, whichever they might be, would survive to adulthood, reproduce, and pass along those genes. For example, those who migrate to Mars will likely lose bone mass in the lower gravity. This leaves everyone susceptible to fractures. In a few generations, though, it may be that those people genetically inclined to have thicker bones have a better chance of survival. Their bones would be thickest among the thin. And over time, perhaps counterintuitively, Martians may become thicker-boned compared with their comrades on Earth because of this natural selection for sturdier bones. Over an even greater length of time, environmentally-induced changes such as larger eyes might get locked into the genome if those with the very largest eyes are deemed the most attractive and fit for reproduction.

Ideally we'd want thousands of people on Mars, not merely for diversification of skills but also diversification of the gene pool. Most modern members of the Amish religious community in Pennsylvania are descendants of the first hundred families to immigrate. As a result of limited genetic diversity and a preference for marrying within their group, the community experiences a higher concentration of genetic conditions, such as dwarfism and numerous metabolic and nerve disorders that go unnamed. Conversely, most other early immigrants to North and South America were soon met with waves of migration, some forced, as in the case of slavery and indentured servitude, which nonetheless helped diversify the gene pool so that genetic disorders were rarely concentrated in one group.

Terraforming Mars

What's the endgame? Some say it is to terraform Mars to make it a little blue marble like Earth. *Red Mars, Green Mars,* and *Blue Mars*—these are the three titles in Kim Stanley Robinson's excellent Mars trilogy, written in the 1990s, in which the author uses science fiction to explore the ramifications of terraforming the red planet. There are biological and engineering considerations, but also philosophical ones. In this sci-fi trilogy, some characters want to keep Mars red and some view greening Mars as spreading life, a gift humans can bring to other worlds. All the while there's a mixed contingency of those who want to break away from Earth, the motherland, à la the American Revolutionary War.

I cannot hide my preference for terraforming Mars. I don't see the ethical concerns. An asteroid could wipe out any planet; and a meteoroid with hitchhiking microbes might have brought life to Earth (or Mars) in the first place, with no consideration for the ethics of it. That is, humans are not unique in our ability to destroy life—or to spread it. I would agree that any life native to Mars should be protected to the best of our ability. But I think our imported life on the surface and immediate subsurface of Mars would not interact with native species. Mars once was more conducive to harboring life a billion years ago, so why would it be wrong to make it that way again, in the same manner that we restore dead land on Earth? Moreover, I feel humans have a moral obligation to spread life in the universe if we can. What's wonderful about a universe if there's no life in it to comprehend or appreciate it?

Philosophy aside, the science of terraforming Mars is worth investigating here. The process would take hundreds of years, at best, although not long in geological time. Making Mars more habitable—for humans, anyway—involves adding more atmospheric pressure, oxygen, heat, or gravity—or any or all combinations. Unfortunately, there's nothing we can do about the low gravity. Among the other three, pressure is the most desired. We can handle cold, and walking about with an oxygen tank isn't horrible. What is horrible is walking around in a pressurized suit and being worried that that suit or even a habitat will lose pressure, killing you instantly. Atmospheric pressure is essentially weight, measured as mass per square inch or centimeter. So, to increase the atmospheric pressure on Mars, we need to put more mass into the air.

One way to do this would be to liberate the carbon dioxide that's frozen at the poles or locked in the soil across Mars. This would put the CO_2 above us in a gaseous form instead of below us in a solid form: same mass; different location. One proposed method for vaporizing the CO_2 is busting the poles apart with nuclear bombs, which I think is absurd. Non-physicist Elon Musk, among other colonization advocates, has discussed this. Although this might be effective in accomplishing the primary goal of increasing atmospheric pressure, given enough bombs, we will have just made Mars a nuclear wasteland—and destroyed the topography of the polar caps and any traces of life the planet may have held. Another method would be to use giant mirrors in orbit around Mars, hundreds of square kilometers across, to focus solar heat on Mars. This would slowly vaporize both dry ice and water ice. Nice, but building such large mirrors would be an engineering feat far beyond anything yet accomplished.

And there are problems to overcome with either approach. First, is there enough CO_2 to create enough pressure? According to a paper published in *Nature Astronomy* in 2018, no.[39] Adding up all known sources of CO_2 on Mars, the researchers calculate that there's enough to increase air pressure about tenfold, but that's still about 90 percent short of what's needed for Earth-like pressure at sea level and about 50 percent short of the low pressure felt on the Earth's high mountains. This is but one analysis, however, and surely not the final word. There may be deep, undetected levels of dry ice in the regolith. Nevertheless, the second problem is that there would then be a thick layer of CO_2 in the air, which is toxic to breathe. Mix with water vapor and you have acid rain to contend with. Also, filling the atmosphere with CO_2 will mean we'll never get rid of the stuff without the millennia-long process of converting CO_2 to O_2 with microbes and then plants.

Yes, a thick CO_2 layer is useful to trap heat. This is a greenhouse gas, after all, and minuscule rises in CO_2 levels on Earth are changing the climate here. So, CO_2 could solve the air pressure and heat problem but not the oxygen problem. The most desired way to thicken the atmosphere would be with oxygen and inert gases such as nitrogen, helium, argon, and neon. That's in the realm of science fiction for now, though. You would need to scoop nitrogen and other gases from Uranus and deliver them via freighter ships numbering in the tens of thousands.

Enter the turtle: slow and steady wins the race. We know how to make worlds warmer; we're doing that right now to our own. Setting up factories on Mars with the sole intent of producing greenhouse gases could slowly but dramatically warm the planet. Among the most potent greenhouse gases are fluorocarbons such as tetrafluoromethane (CF_4) and hexafluoroethane (C_2F_6), which linger in the atmosphere for tens of thousands of years.[40] Those notorious chlorofluorocarbons, CFCs, would be good, too, but they would destroy ozone, which is needed for UV radiation protection. Fluorine and carbon both exist on Mars, and creating factories to add fluorocarbons to the atmosphere in parts-per-million levels could warm the planet two degrees per decade. In two decades, a four-degree rise could cause a runaway greenhouse effect, liberating the dry ice at the poles so that, in a hundred years, all of the CO_2 frozen at the poles would be up in the air.[41] Chris McKay and his colleagues further calculated, based on laboratory experimentation, that the most effective combination of greenhouse gases in terms of longevity and ability to trap overlapping frequency bands of solar radiation would be CF_4, C_2F_6, C_3F_8 (octafluoropropane), and SF_6 (sulfur hexafluoride).[42]

Feel the Pressure

With each bit of frozen CO_2 melted, the atmospheric pressure rises. Like rolling a large stone downhill, there's no stopping this process once it starts. However, the high-end estimates predict there's only enough CO_2 on Mars, at best, to create 350 millibars of pressure, about a third of that on Earth. So, there won't be too much CO_2 to cause too much air pressure, like the situation on Venus. And even 150 millibars is livable without a thick pressure suit. The Armstrong limit, the lowest pressure before blood begins to "boil" at room temperature, is about 63 millibars. Any level below 63 millibars, which is the air pressure at an altitude of eighteen kilometers above Earth, requires humans to wear a pressurized suit. But unprotected at, say, 64 millibars is not pleasant. Your surface fluids such as tears and body moisture will evaporate, and you cannot effectively transfer oxygen from your lungs into bloodstream and saturate hemoglobin. For normal bodily functions, humans need an air pressure of about 150 millibars while breathing 100 percent oxygen.

If our goal is to make Mars warm, with enough atmospheric pressure to walk about without a pressure suit, just with oxygen, let's examine Mount Everest. On the summit of Earth's highest mountain, at 8.850 meters, the air pressure is about 340 millibars. We think that's attainable on Mars. And at 340 millibars, your bodily fluids will be fine. The only concern is the ability for oxygen to saturate hemoglobin in the blood at that pressure. Levels below 90 percent saturation results in hypoxemia and can cause tissue damage, as cells are deprived of oxygen. That's the primary cause of death on Mount Everest. Climbers need supplemental oxygen not just because the air is thin but also because the atmospheric pressure is not enough to push the oxygen into their lungs and allow for the proper exchange of CO_2 for O_2. The O_2 concentration on Everest is about the same as it is at sea level, about 20 percent; but climbers sometimes breathe from tanks containing 100 percent oxygen to ensure they can absorb the maximum because of the poor exchange of gases at this low level of air pressure.

Studying people who live at high altitude on Earth can help us understand the possibility of life on Mars in terms of higher radiation exposure and lower atmospheric pressure. Tens of thousands of people live above at altitude of 4,700 meters, where the atmospheric pressure is about 550 millibars. They seem to have no higher incidence of radiation-induced cancer. Also, these people are acclimated to the thin air, able to breathe without assistance.

Thus, we have real-world experience to suggest that raising atmospheric pressure on Mars to at least 350 millibars, Everest level, will enable humans to walk around without pressurized suits while breathing pure oxygen; but raising pressure to 550 millibars will ensure that humans can get by on a "normal," 20 percent oxygen, air mix. With a thicker atmosphere comes better radiation protection.

That's for humans. Bacteria, fungi, lichen, algae, and plants can tolerate lower pressures, close to the Armstrong limit for the simplest of organisms. At an atmospheric pressure of a few hundred millibars and temperatures above freezing, we could expect an Alpine biome—someday. Regardless of the warmth, few plants can survive well in 95 percent carbon dioxide, which is what Mars is. To add organisms to Mars, we need to understand how life started on Earth. Billions of years ago, the atmosphere was mostly nitrogen, carbon dioxide, and water, with no oxygen gas. The leading theory is that, in the presence of liquid water and sunlight, cyanobacteria converted CO_2 to O_2 so well that most of the carbon dioxide was scrubbed from the atmosphere to a point of quasi-equilibrium, today, of about 0.5 percent CO_2 and 20 percent O_2. That could work on a slightly warmer and higher-pressure Mars.

Although it would take at least 100,000 years to raise oxygen levels to 20 percent for humans, seeding a warmer Mars with cyanobacteria alone could produce enough oxygen for higher-level organisms such as lichen and some primitive plants to appear in a hundred years. Remember that habitability doesn't mean perfection. Think of it in terms of edibility. There's food that's edible, and there's food that's delicious. Mars may someday become delicious, but it will take time to cook. In the meantime, Mars would be increasingly easier to digest with each degree of heat, millibar of pressure, and percentage of oxygen added to the dish.

An interesting idea explored by Kim Stanley Robinson in his Mars trilogy is the creation of moholes, a word stemming from the Mohorovičić discontinuity, the boundary layer between the Earth's—or a planet's—crust and the mantle. In short, moholes are created using automated, cylindrical boring machines to dig wide holes tens of kilometers deep to reach the boundary layer. On Mars, they would be used to release heat and promote global warming. But because the holes are so deep, humans could live at the bottom of them and experience Earth-like pressure. Moholes have been attempted on Earth, although not successfully.

I view digging moholes on Mars as a solution somewhere between fluorocarbon burning and Uranus gas imports in terms of feasibility. Another idea is to skim comets or asteroids across the upper atmosphere in a controlled maneuver to burn off volatiles—water, ammonia, oxygen—to add valuable gases to Mars. This idea is less technologically challenging then importing gas in space-tankers; a few big rocks will go a long way, as opposed to thousands of deliveries from Jupiter. The math better be perfect, though, or those massive objects could kill the settlers if they hit the surface. Bombarding the surface deliberately would work equally well for providing atmospheric gases, but this method is crude and dangerous and runs counter to the notion of conducting science and establishing settlements on Mars.

It is true that Mars lost its original atmosphere from a combination of low gravity and lack of magnetosphere, which enabled solar radiation to blow the atmosphere away into space, but the process took tens of millions of years. So, we'd have some time to enjoy on Mars before it becomes uninhabitable again.

The Future of Mars

Terraforming Mars is not necessarily making a new Mars. One can view this process as bringing out the old Mars, a return of flowing water and, perhaps, ancient Martian life. And here's where the opinions of many Mars advocates diverge. If there is life on Mars, it must be closely examined. If it is like our tree of life, based on RNA and DNA, then it may very well be related—either Mars seeded life on Earth, or vice versa, long ago. If we are of the same tree of life, then there should be no ethical reason why Earth life cannot be cultivated on Mars. If the life is utterly different, however, if it represents a second or independent genesis, then some terraforming advocates such as Chris McKay argue that it is humanity's obligation to cultivate that Martian life remotely and to keep off the planet altogether. Indeed, this is where McKay splits with Zubrin—and me, I guess. How about you?

My prediction: A full-blown space race between China and the United States by the 2030s; first Chinese-led or US-led mission by the 2040s; multination, permanent presence on Mars by the 2050s, including some tourism; first attempt at a nongovernment settlement by the 2080s; cyanobacteria, lichens, fungi, and simple plants growing naturally by 2150; a Mars of Icelandic temperatures and livable atmospheric pressure by the twenty-third century.

7

Living in the Inner and Outer Solar System and Beyond

The only concept more awesome than the immense scale of the observable universe—billions of light-years across, with at least one hundred billion galaxies, most with billions of stars—is the fact that the moist, squishy little object just a few centimeters across called the human brain can begin to comprehend it.

Allow me to exercise that brain for a moment to visualize the distance from Earth to the other planets, all mere specks in our galaxy. There is no way to properly relay the scale of the Solar System on a standard sheet of paper. Once you draw the Sun, you can't render Earth to scale. A million Earths could fit in the Sun. You might render Earth as a dot, but that dot would be at the opposite end of the page. You haven't even gotten to Mars and Jupiter, let alone Pluto. On the National Mall, in Washington, DC, sits a one-to-ten billion scale model of the universe, called Voyage. It stretches for more than six football fields, four city blocks. The Sun is about the size of a large grapefruit. Mercury, the closest planet to the Sun, is nine steps away, or about six meters; Venus is an additional eight steps away. Then Earth is about six more steps and is about the size of a pinhead, or a segment of an ant, with the

Moon a fleck of matter directly next to it. This is the entirety of direct human experience, fifteen meters from the Sun, just a fleck and a pinhead. Mars is a bit farther out, twelve steps. One can hardly see the model of Mars, as it is so tiny compared with the grapefruit Sun, albeit three times as large as the Moon speck.

Now the distances start getting larger. Jupiter is 50 meters down the road from Mars, five times more distant from the Sun than Earth is. About 900 Jupiters could fit into the Sun, and the Jupiter model is the size of a pellet. Saturn is about 65 meters down the road from Jupiter, a slightly smaller pellet; Uranus is another 140 meters; and Neptune 160 meters more. Pluto, at 135 meters more, is so small compared to the Sun that is too tiny to see at this scale; 250 million of them could fit in the Sun. That's the end of the Voyage model but not the end of the Solar System. The Kuiper belt would be yet another 150 meters down the road, and interstellar space would be 750 meters more. Then the coast is clear. Proxima Centauri, the closet star, at this scale, would be in California and the size of a cherry. You can begin to understand why it takes months to reach Mars and five years to reach Jupiter. NASA's *New Horizons,* one of the fastest crafts ever flown, took nearly ten years to get to Pluto.

One must appreciate this scale as we discuss the feasibility of colonizing worlds beyond Mars and the asteroid belt. Humans may visit the moons of Jupiter and Saturn this century, but a far longer length of time would be needed before we could ever establish a scientific base or settlement in the outer Solar System. Missions take time to develop, and we need reconnaissance before sending people. The Soviets and Americans combined attempted or succeeded in sending a staggering seventy-one missions to the Moon before successfully landing humans there. Similarly, to date, there have been fifty-six mission attempts to Mars, about half of them

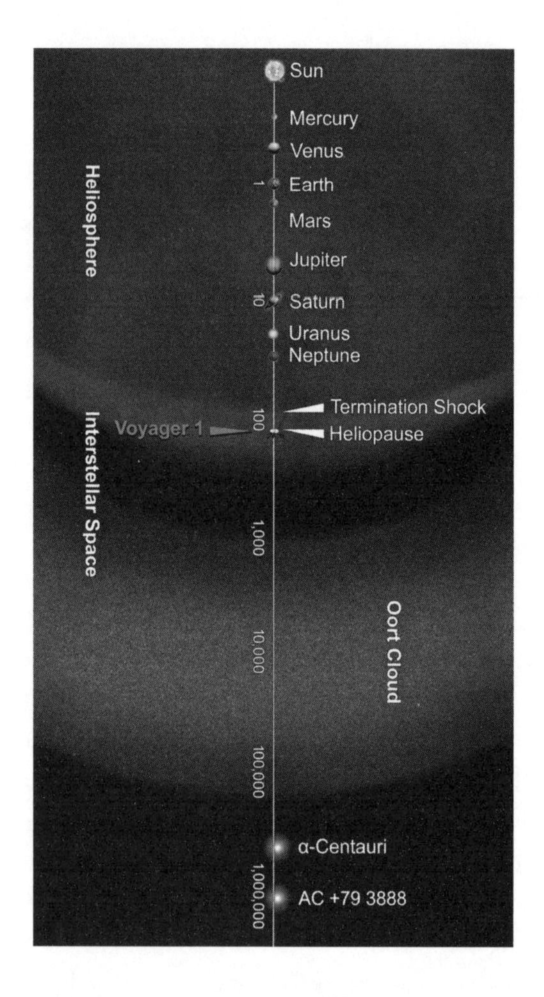

This chart places the Solar System distances in perspective. The scale bar is logarithmic and in astronomical units (AU), with each marked distance beyond 1 AU representing ten times the previous distance. One AU is the distance from the Sun to the Earth, which is about 93 million miles, or 150 million kilometers. Neptune, the most distant planet from the Sun, is about 30 AU. The most distant human object is the satellite Voyager 1, launched in 1977 and more than 145 AU from Earth as of 2019.

failures. With Jupiter and beyond, it becomes more difficult to send many missions because of the combined time needed to develop missions and reach the destination.

The New Horizons mission concept dates back to 1992; was selected in 2001; was canceled in 2002; was reselected in 2003; was launched in 2006; and reached Pluto in 2015. From initial assembly to finish line, that was at least fourteen years. The fastest we could do missions to the outer Solar System is ten years: five in development, five in flight. In June 2019, NASA announced a very cool mission to send a rotorcraft-lander called *Dragonfly* to the Saturn moon Titan, which will launch in 2026 and arrive in 2034. We would likely need to send several robotic missions to any given planet before sending humans. This time limitation, plus competing priorities—establishing bases on the Moon, bases on Mars, bases on asteroids—extends the timeline for creating deep-space colonies.

Consider the Mars example: the logical plan is to send supplies first; confirm that all is working; and then send humans. For Jupiter, it would take five years for reconnaissance or supplies to reach the desired moon, so any human mission would inevitably be "delayed" for that long to ensure everything is feasible and working before setting off toward the destination. Even turning toward the Sun, the relatively nearby inner planets present their own challenges of tricky orbital maneuvering near our massive star.

In this chapter, I will discuss life at the extremes, first dipping sunward toward Venus and Mercury, then pulling out to Jupiter and beyond. While fascinating landing sites can be found, the requirement for safe harbor in these blistering hot or blistering cold regions would essentially require the same technology: solid habitats that can withstand temperatures far beyond the realm of human experience.

Cloud Cities of Venus

Venus is both the worst and best place in the Solar System to establish an off-Earth colony. Often called Earth's sister planet, Venus is nearly the same size and mass as Earth and offers 0.9G. That level of gravity is almost surely enough to guarantee normal growth and development for any human living on Venus—a huge plus. But there's an ugly side of Venus, which you know all too well. The average and uniform surface temperature is 465°C (870°F), making it the hottest in the Solar System, even hotter than Mercury.[1] That's hot enough to melt lead. We could theoretically create structures with metals with higher melting points—steel, iron, or nickel—but we would still have to deal with a crushing atmospheric pressure that's ninety-three times greater than on Earth's surface.

The first two Soviet probes sent to Venus were crushed liked soda cans before even reaching the surface; the next two landers, Venera 7 and Venera 8, touched down and transmitted data for a full hour before succumbing to the pressure.

Venus is this hot not simply because of its proximity to the Sun but also because of its thick atmosphere of carbon dioxide. Several billion years ago, Venus was probably more like Earth, with a liquid ocean and conditions suitable for life to evolve. Then something happened. Planetary scientists don't know what, exactly. But something—perhaps a huge asteroid impact—liberated water and CO_2 on or under the surface. This triggered a runaway greenhouse effect: incoming heat from the Sun was trapped, heating the surface more and more until most of the CO_2 that was once underfoot was now overhead. Water went up, too, got heated, got stripped apart into hydrogen and oxygen, and was lost forever.

The surface of Venus is worth exploring but not living on. Sound and vision are distorted; the dim light bends in this pressurized world, and the sky is not visible through the 15 km layer of clouds. We do have an interesting option, though, of living in those clouds. This sounds futuristic but is absolutely feasible. Venus's atmosphere is so thick that we could almost sit on top of it. Because oxygen and nitrogen are lighter than CO_2, humans could live in massive balloon bubbles filled with these gases at a height of about 50 km above the Venusian surface, where the temperature is a manageable 50°C (120°F) and the atmospheric pressure would be 15 psi, just like on Earth at sea level.* There would still be enough atmosphere above us to protect against solar and cosmic radiation. Venus is close, too, and we could get there in about three months, or half the time for travel to Mars.

The Venusian wind in the middle atmosphere is strong, upward of 340 km/h (210 mph), which is challenging but perhaps manageable from a balloon-stability perspective.[2] That speed would take you around Venus in about four Earth days. That presents an interesting day–night cycle. Venus spins so slowly that a Venusian solar day is 116.75 Earth days. So, if you were on the surface of Venus, you'd have to wait nearly four months for the Sun to rise, set, and return to the same point in the sky. But in a balloon city, you would be moving faster than the ground below and would experience about forty-eight hours of sunlight followed by forty-eight hours of darkness. With some energy exertion, the balloon could travel at a speed that maintains its position in the sunlight. One more perk is that you could leave the confines of your balloon city to step outside on an exposed platform, provided you have an oxygen tank. The pressure and temperature would be just fine. You could hang glide, if you dare. Just don't go too low.

*There's a trade-off at 50 km, of cooler temperature at slightly higher altitudes but precipitously lower pressure.

One challenge would be dealing with the sulfuric acid clouds. Sorry, forgot to mention that part. The sulfuric acid could eat away at the balloon city, sending it tumbling to the fiery hell below. We would need a special material such as polytetrafluoroethylene that can repel sulfuric acid. We'd need a lot of it, too, for the city-size balloon and for bodysuits to venture outside. We'd also need to bring a water supply, which is feasible.

There's only one major catch: Why do this? Colonizing Venus in cloud cities is a prime example of a path that's technically feasible, and thus promoted by some futurists, but bereft of practicality. What would draw millions of people to live in the clouds of Venus, forever, with no view of the world below and only a cloudy view of the star field above? The thrill of space could be provided in human-made orbiting cities between the Earth and the Moon, all with suitable artificial gravity, pressure, and temperature, yet with easier accessibility and better scenery. Similarly, what would drive the economy of Venus? Whatever that might be, it is not obvious. So, government or commercial investment and subsequent establishment of bases and colonies would first favor other options presented in this book, namely the Moon, Mars, asteroids, and orbiting cities.

Terraforming Venus is a possibility for some future, technologically advanced civilization. The one advantage is that, because no one will be living there, unlike on the Moon or Mars, you can destroy the hell out of it (literally) to make it livable. The concepts are bit too far-out to discuss in detail here, but the gist is to bombard the planet with an icy moon to add copious amounts of hydrogen, which could react with CO_2 to form graphite and water, which would fall out of the sky and leave a much thinner atmosphere of about 45 psi.[3] A similar, violent idea is to bombard the planet with magnesium and calcium to create oxides that, again, would fall from

An artist's impression of balloon cities above Venus. The weather on Venus isn't so hellish if you stay up in the clouds. Balloon cities filled with oxygen and nitrogen could float above Venus's thick carbon dioxide atmosphere. The temperature, pressure, and gravity level would be similar to Earth. NASA is entertaining the concept with the proposed High Altitude Venus Operational Concept (HAVOC) project.

the sky and take carbon with them.[4] Carl Sagan had a gentler idea, circa 1961, of seeding the atmosphere with bacteria to eat the CO_2; but this was before we understood the density of the CO_2, and Sagan conceded in the 1980s that this plan wouldn't work.

Rail Cities of Mercury

You know how, when you're walking on a hot beach, you have to keep on moving so your feet don't burn? That's how it is on the planet Mercury. Move or melt. On Mercury, you shouldn't keep still.

Mercury can get almost as hot as Venus. But the difference is that, at night, it can dip down to −170°C (−280°F). So, the average temperature of the planet is much lower. The reason is that there's no atmosphere to trap and circulate heat, just like on the Moon. Mercury also has permanently shaded craters that are among the coldest regions in the Solar System. In fact, Mercury is very much like the Moon in terms of size and terrain, pocked with crater impacts. I mentioned that the bearable times on the Moon are during the multiday dawn and dusk, as the temperatures then are between the insanely cold and insanely hot. Something similar applies to Mercury. The planet spins rather slowly, and a full day is 175 Earth days. In fact, its solar day is longer than its solar year, which is eighty-eight Earth days. As a result, there is the phenomenon of a near-perpetual sunset, in which the terminator line—the division between day and night—is moving at about 3.5 km / hour. On Earth, the line moves about 1,600 km / hour, and you'd need to be in a jet plane to experience a perpetual sunset.* On Mercury, you could experience this, and the comfortable temperature it provides, on a slow-moving train around the circumference of the planet.

A habitat on rails running westward could stay in the cool zone.† Explorers could, in theory, walk on the planet in these zones, provided they have the same pressure and oxygen supply as they would have on the Moon. Otherwise, they could live underground, preferably near a shaded crater that might contain water, and come out during the extended local dawn or dusk to explore. Similarly, as on a hot beach, we could build massive, reflective umbrellas to live under. One very neat feature of being on Mercury would be to see the Sun move backward, temporarily. This is an optical

*British singer Roy Harper has a beautiful song about having this experience while flying, called "Twelve Hours of Sunset."

†Break down and you're toast—or ice.

illusion, as neither Mercury nor the Sun is moving in reverse. But the speed of Mercury's orbital motion around the Sun is so fast compared with its own spin that, as a result, the Mercurian visitor would see the Sun rise partway, reverse, then set back in that same spot it rose from, and then rise again—a double sunrise.

We call Mercury a rocky planet, like Earth, Venus, and Mars, but it is more properly classified as a metallic planet: about 70 percent metal (mostly iron and nickel) and 30 percent silicate material. Indeed, Mercury is so dense with metal that, although it is only slightly larger than the Moon, it has a gravitational pull as high as Mars's gravity, 0.38G. Now, why anyone would want to live on Mercury is beyond me. Yes, there's Martian-like gravity that might be suitable for life, but at the price of living underground or on a slow-motion train-city with limited ability to explore and enjoy the environment. Life on Mars would be easier and less dangerous in this regard. Mercury does have a modest magnetic field that blocks some solar radiation, but you will get blasted. There's limitless solar energy to power heavy industry on Mercury, and one could envision mining operations there. But mining asteroids would be easier and closer to potential markets. Mercury also is very difficult to land on and leave from. The planet is moving very fast, and adjusting to the precise delta-v requires not only much fuel but also incredible finesse. If you're off by just a hair, you will go tumbling into the Sun.

Terraforming Mercury seems out of the question, too, because there's no atmosphere to even start with; all volatiles would need to be imported from the outer Solar System; and any protection you created likely would get undone by the solar wind without some kind of planet-wide heat shield. One solution could be to move the entire planet away from the Sun with directional, fusion-powered explosions. We would have to *really, really, really* need the metals or land to invest in such an undertaking.

Life on the Jovian and Saturnian Moons

It's a hike to get to Jupiter and Saturn, but should we have the technology to send humans to these gas giants, there's much to explore. These planets themselves have no known surface and offer little opportunity to live on or near them. But their moons can offer a limited amount of safe harbor, although none with gravity much greater than our own Moon.

Jupiter has at least seventy-nine moons. One of these, Callisto, is the most habitable from a human perspective, although the moon Europa appears to have a liquid ocean below thick ice that could harbor alien life and that is certainly worth visiting. Similarly, Saturn has at least sixty-two moons. The moon Titan has an atmosphere thick enough to provide suitable pressure and radiation protection. There are even flowing rivers, lakes, clouds, and rain—a complete cycle, albeit constituting liquid methane and ethane, not water. A human habitat here is remarkably feasible. The moon Enceladus has a sub-ice liquid ocean that occasionally sends plumes of water vapor into the air, another candidate for alien life.

Living in these extremes is pretty much all the same. Cold is cold, dark is dark, distant is distant. I can see little difference in the challenges of living around Jupiter or Saturn compared with living on Pluto. To make the five-year journey to Jupiter, one must assume we have reached a stage in which large, comfortable, and protective spacecraft are available to propel us anywhere in the Solar System in ten to twenty years. With safety assured, the top issue is the time commitment, or the years that an explorer or migrant must allot to reach these distant locations. Let's start with Jupiter, likely our first planetary destination after establishing bases on Mars.

Jupiter and the Galilean Moons

Jupiter is enormous, more than twice as large as all the planets and moons in Solar System combined. Like the Sun, Jupiter is a big ball of hydrogen. Jupiter, however, does not have the mass to generate the temperature and pressure to initiate hydrogen fusion. For that, Jupiter would need to be at least seventy-five times more massive. So, to call Jupiter a failed star, as some have, is a bit of a stretch. Jupiter didn't fail. If the planet sucked up all the other non-solar mass in the Solar System, it still wouldn't have been close to being a star. Jupiter, instead, should be seen as a mighty and unique planet.

Concepts for colonizing Jupiter itself, not the moons, are squarely in the realm of science fiction. There's no surface. And with an atmosphere that's 75 percent hydrogen and 24 percent helium, the two lightest elements, there's no way to float above it. A helium balloon would sink like a lead balloon. Worse for life, Jupiter's massive magnetic field catches particles from the Sun like a net and whips them about to even higher energies, showering that deadly radiation across many of its moons. Jupiter is fascinating; its core is likely metallic hydrogen and it may rain diamonds on the planet. Yes, diamonds. But there's simply no practical way or reason to live on Jupiter to explore these phenomena up close.*

The four larger, inner moons of Jupiter—Io, Europa, Ganymede, and Callisto—are of keen interest. We call these the Galilean

*Futurists speak of living on orbital rings and mining Jupiter for hydrogen to fuel the fusion economy, all fascinating and feasible, and all centuries off into the future.

moons because they were discovered by Galileo in 1610, the first objects found to orbit another planet. You can see these moons, too, with a simple pair of binoculars. Io is the closest to Jupiter and is slightly more massive than our Moon; it has more than 400 active volcanoes, the most geologically active moon or planet in the Solar System. The churning of its insides seems to be caused by gravitational tidal heating as a result of being so close to massive Jupiter. We could land on Io. But the thin atmosphere of sulfur dioxide and near-zero atmospheric pressure would make for an unpleasant stay. Also, Io is dosed with 3,600 rem per day, which would kill any human swiftly.[5] Apologies to the geologists and volcanologists, but Io is a place to send robots, not humans.

Next up is Europa, slightly less massive than our Moon. This moon is famous for its probable liquid ocean below kilometers of ice. Visit Europa? Maybe, but only quickly. Radiation there is 540 rem / day.[6] Interestingly, Europa has an oxygen atmosphere, but it's about a billion times thinner than Earth's, barely detectable. That's OK, because all the action is below the ice. Conceivably, we could establish a highly fortified scientific base on the ice akin to an ice-fishing hut that could shelter us from the intense radiation and the −160°C temperature at the equator, surprisingly warm given the distance from the Sun. (That heat is from Jupiter's tidal effect.) From this scientific base, we could bore through the ice to the ocean below. Presumably, by this point, we would have sent robotic missions exploring this ocean to see whether life there is even possible; that is, our human presence would be a follow-up to explore alien life already discovered or strongly hinted at.

The biggest obstacle would be to bore through that ice, which is as hard as granite. We don't know how thick it is; estimates are between 10 and 100 kilometers.[7] For comparison, Lake Vostok in Antarctica is covered with about 3 kilometers of ice. Nuclear power would be needed to bore through all that ice. Then, a submersible could be deployed into this ocean, estimated to contain two to three times more water than Earth. Europa is such a strong contender for life not just because of the liquid water but also because of the possible hydrothermal activity, similar to undersea volcanoes on Earth, only here invigorated by the tidal tug of Jupiter. If life could exist in oceans on Earth fed by hydrothermal vents, not sunlight, then life could survive on Europa. The core question is whether life could *originate* on Europa in the first place; or are conditions of sunbaked wet and dry areas, like tidal pools on Earth, needed for a genesis? The ice does provide protection from radiation and the necessary atmospheric pressure for life.

I adore Europa. But given the dangers, I can't imagine a situation in which a submarine controlled by humans in situ under the ice would be better than one controlled by humans remotely from the surface or, better yet, from the relative safety of the moon Callisto. On Mars, hands-on contact is preferable. On Europa, purely robotic exploration might be superior. Similarly, what benefit a settlement or colony on Europa would bring is beyond me. People there could perhaps operate the tourist trade, taking visitors below the ice to view alien life. Lots of assumptions here: life discovered; it's visible and majestic; the ocean is safely accessible; rules permit human contact with alien life; and so on. Environmental dangers aside, the surface gravity of 0.13G would seem to rule out any long-term human presence.

Next up is Ganymede, Jupiter's largest moon, larger than our Moon but less dense, so the surface gravity is only 0.15G. Same deal: frigid, nearly airless, and bombarded by Jupiter's radiation at a rate of 8 rem / day (compare with Mars at 8 rem / year).[8] Ganymede seems to have a lot of water ice and perhaps a sub-ice liquid ocean, although the case for that is not as solid as it is for Europa. There's ice-mining potential here, depending on how thirsty future generations are and whether the asteroids can't slake that thirst. Again, there seems to be no compelling reason why anyone could call this moon their home, given the challenges of life there and better options elsewhere.

This leaves Callisto, Jupiter's second-largest moon. The best feature of Callisto, compared with the other large Jovian moons in terms of human habitability, is the relatively low radiation levels, only about 0.01 rem / day, ten times higher than what we're exposed to on Earth but still less than on Mars.[9] With the other moons equally cold and airless and lacking suitable gravity, Callisto is the winner among the less-than-ideal pack. From a base on Callisto, we could visit the other large moons in a day or so. Communication to these moons would be nearly instant depending on their orbital positions relative to each other and the number of communication satellites we have in the region. The regolith on Callisto is like gravel, and excavation in the low gravity should not be difficult. As with Ganymede and Europa, Callisto has water ice and perhaps liquid water beneath this; and that could supply water for drinking, farming, breathing, and burning. Like our Moon and most of Jupiter's moons, Callisto is tidally locked to its planet and shows only one face to Jupiter. That's the side to be on. On Callisto, Jupiter appears five times as large as a full moon on Earth; what a sight that would be. The moon has a seventeen-day / night cycle, shorter than our Moon's twenty-eight days.

Callisto could be the logical place to set up camp with the goal of exploring all the Galilean moons in relative safety if an orbiting space hub in that vicinity couldn't serve this purpose. In 2003, NASA conceptualized a human journey to Jupiter in the 2040s called the Human Outer Planets Exploration (HOPE) mission, with an eye on Callisto as a landing site. Talk about hope. We'd be lucky to get to Mars by that time frame. Nevertheless, the mission architecture highlights Callisto as the likely place for human presence in the Jovian system.

As for a permanent human settlement near Jupiter, we see the conversation dominated by terrestrial chauvinism. The artists draw illustrations of bases on the lunar surfaces. But why live on any of these moons, even Callisto? What benefit does solid ground offer? Any long-term presence on these moons would require shelters that are underground or otherwise encased in a structure that would divorce you from the experience of living on that moon. Future generations who want to live near Jupiter—for the brilliant view, for the opportunity study this planet and its moons, or to coordinate some commercial enterprise—would likely benefit more from living in a rotating, orbiting human-made city or massive space hub with artificial gravity. All points of interest would be in immediate proximity, should you want to visit. Unless 0.15G is somehow a good thing for human health, humans in the twenty-second century would choose to live in the comfort of artificial gravity.

Saturn and the Calling of the Mighty Titan

Saturn, famed for its rings, is almost a third the mass of Jupiter; and like its bigger brother, the planet is mostly hydrogen and helium. Saturn has dozens of moons, many unnamed, as well as hundreds of moonlets a few tens of meters across lodged in its icy rings. Many of these moons have fascinating features. For example, tiny Tethys is almost entirely water ice, a perfectly round ice ball 1,000 kilometers in diameter. Mimas is the smallest astronomical body known to be a near-perfect sphere as a result of self-gravitation; Mimas is about 400 kilometers across and, with its large crater impact, resembles the Death Star from the movie *Star Wars*. Iapetus has a strange equatorial ridge that makes the moon look like a walnut from certain angles. And Rhea—who could forget Rhea—is like a dirty snowball of silicates and ice some 1,500 kilometers in diameter.

But two Saturnian moons take center stage: diminutive Enceladus and the mighty Titan. Enceladus is only 500 kilometers across, smaller than the asteroid Vesta, with no atmosphere and a surface gravity of 0.011G, the lowest of any object I have discussed in this book other than Mimas. What draws us to Enceladus is the search for life. The moon contains a subsurface ocean that regularly erupts in plumes of water vapor over its south pole. That water vapor actually forms part of Saturn's ring system. From observations with NASA's Cassini spacecraft, we also know that the ejecta contains all the indirect signs of life: salts, ammonia, silica, numerous organic molecules such as methane and formaldehyde, and hydrogen gas.[10] Among these, hydrogen gas is particularly important, as it is a sign of hydrothermal vents—that is, food.[11] Such compelling evidence for the possibility of finding alien life pushes us to Enceladus; and several missions are being designed to fly by, scoop up a sample from those plumes, and bring it back to Earth.

Living on Enceladus would be identical to living on Europa, in an ice-fishing hut. The gravity is lower, but the radiation is less intense. Still cold, dim, and devoid of air, though. Titan in some ways is a more interesting place to live. This is the largest of the Saturnian moons—the second largest in the Solar System, behind Ganymede. And along with Ganymede, Titan holds the distinction of being a moon that's larger than the planet Mercury. Titan is the only moon to have a dense atmosphere, 1.4 times thicker than Earth's, in fact. Also, Titan is the only other known body in the Solar System that currently has a surface liquid that flows in rivers and lakes and rains down from clouds. That liquid is methane and ethane at a cool −180°C.

Because of its atmosphere solving the radiation and pressure problems, some people suggest that Titan is the best place in the Solar System to call our second home, a location even more choice than Mars.[12] Slow down, I say. That's a ludicrous idea. We may get to Titan, but certainly not before Mars. And there's only a slim chance we could live there.

The first challenge is the distance. Titan is about 1.4 billion kilometers from Earth, about twenty-five times the distance to Mars. It took the ESA Huygens mission six years to get to Titan.* The mission was a fabulous success. ESA actually placed a probe on the surface of Titan in 2005, the most distant probe ever landed. The video of the graceful descent is stunning, providing a panoramic view of the moon's surface below the clouds, our very first peek; and the probe collected data for a full ninety minutes before going quiet. ESA demonstrated the feasibility of landing a craft of Titan, despite the eighty-minute communication gap (radio waves traveling at light speed, if that's any indication of how far that is). Nonetheless, we would need a rather large spacecraft to send humans there in comfort and good health. The second challenge is the low gravity, 0.14G, less than Earth's Moon and less than half that of Mars. So once again, would-be space settlers need to pick their poison: deal with the radiation and pressure issues on Mars, which can be corrected; or deal with the gravity problem on Titan, which can never be corrected.

Let's set aside the gravity concern for the moment. Admittedly, what is compelling about Titan is the freedom to walk around

*Huygens piggybacked on NASA's Cassini probe before descending onto Titan.

without a pressure suit. The world will look somewhat familiar, too, with rain and rivers and clouds and such. You could go boating. You could hang glide rather easily in the heavy atmosphere, dominated by more than 95 percent nitrogen. Indeed, flying would be easier than walking, and preferred. The low gravity and high pressure make for awkward strides, like walking underwater. With simple wings attached to your arms, you could just fly. Or, instead of riding a bicycle, you could travel by "flycycle."* Standing on Titan will be a problem because the heat of your body would melt the ground, which would then refreeze around your feet, as if you're stuck in muck. Flying and treading only slightly and quickly on the ground will be your best bet. In terms of protection, though, you're not out of the cold yet, so to speak. The cold is below anything we experience on Earth, −180°C. The lowest natural temperature ever recorded, on Antarctica, was −89.2°C. To protect yourself from the cold, you might as well be in a pressure suit unless some new "space age" fabric is created to keep you warm without the clumsy bulk. Regardless, every single part of your body must be protected or else any flesh exposed would freeze instantly. Think ultra-insulated scuba gear.

You'll need oxygen, too. This wouldn't be hard to secure. There's no atmospheric oxygen, but there appears to be copious amounts of water ice on Titan, albeit as hard as rock. That is, the rocks we see on Titan might be solid ice. This could be melted to drink and grow crops and, as we've seen, split apart to produce oxygen gas, O_2. With so much nitrogen on Titan, any habitat could easily have an Earth-like mix of 80 percent N_2 and 20 percent O_2 to breath. Fuel also would be nearly unlimited. It rains from

*Note to self: trademark the term "flycycle" by the end of the century, before we get to Titan.

the sky, after all. You might be worried that those methane and ethane lakes would go up in an immense conflagration. But remember, hydrocarbons won't burn without the presence of O_2, and you could easily control that.

First, though, there's a little catch-22 to overcome: you can't burn hydrocarbons without oxygen, but you can't melt water and get oxygen without heat. Solar won't cut it. Titan receives only about 1 percent of the sunlight that hits Earth, and much of that is absorbed by the atmosphere. So, you would need a small nuclear reactor to make the oxygen to burn the methane and ethane. That's why I wrote a "nearly" unlimited supply of fuel. You'd need nuclear—radioactivity, fission, or fusion—to perpetuate the energy cycle, a catch many Titan advocates have glossed over. Nevertheless, Titan could provide methane rocket fuel for export. This moon could be the OPEC of the outer Solar System, if not the entire Solar System, particularly if we never attain our fusion dreams. Hydrocarbon mining, with technology as trivial as using a vacuum machine, could drive a profitable economy to support a human settlement.

Life on Titan would be funky, for sure. Recreational boating remains an option, although the liquid methane–ethane mix is less dense than water, and to provide the necessary buoyancy to float, ships would need deeper and more hollow hulls. Then again, the liquid hydrocarbon is also less viscous than water, and your vessel would slice through the sea or lake with less drag. The wind is tame, but the air is thicker, so there's more stuff to puff the sails. And sails would be in order. Paddling and rowing would difficult in the low gravity and low viscosity. A chemical-propulsion engine in a methane-ethane lake might be a tad dangerous. Swimming could be fun, once you overcome the ick factor of submerging yourself in this gasoline-like soup; its lower density compared with

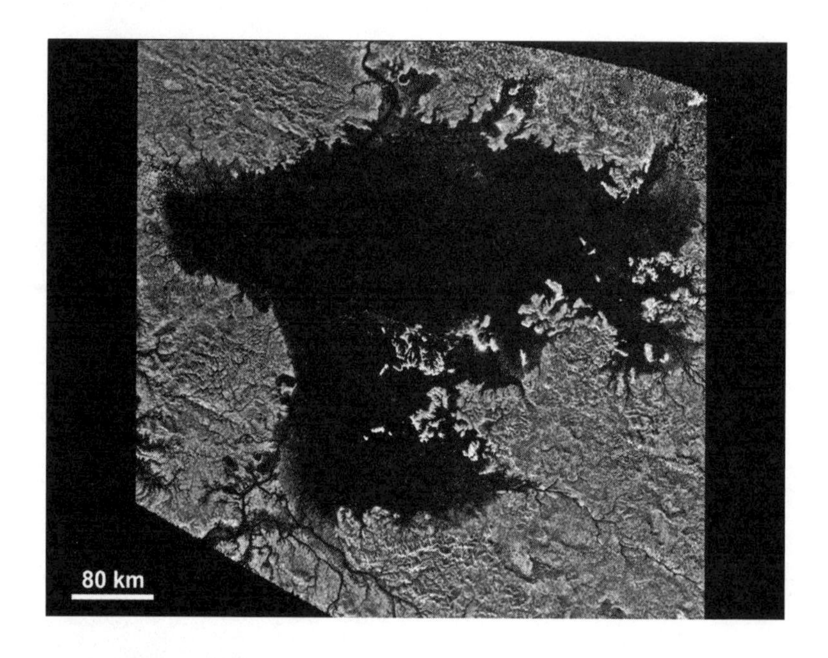

A trip to the lake district. Ligeia Mare is the second largest known body of liquid on Saturn's moon Titan. That liquid is ethane and methane, flowing at –179 degrees Celsius. Titan has a liquid cycle of lakes, rivers, clouds, and rain, and astrobiologists have speculated that life could theoretically evolve in such an environment.

water and the low gravity implies that you may be able to leap out of the liquid like a dolphin, no doubt with a little practice.

Lakes and seas appear throughout Titan, and we have already given names to many: Kraken Mare is the largest, some 400,000 km^2, more than twice as large as the Great Lakes in North America; Ligeia Mare and Punga Mare are two more lakes of scientific and yachting interest. The Titan Mare Explorer (TiME) is a proposed NASA/ESA mission to land in Ligeia Mare and motor about with a nuclear engine. TiME has been passed over for funding, well, time and time again, most recently in 2012. The mission is absolutely feasible, though, as the Huygens mission has paved the way. And the low-gravity, thick-atmosphere environment make for much easier landings compared with Mars or even Earth. This is just a matter of competing priorities for missions to other planets and moons. A follow-up mission to TiME would be a submarine on Titan, far more complicated. NASA has green-lighted a mission called Dragonfly, as mentioned, with a drone-like probe that can analyze air and regolith samples from dozens of sites.

Habitats *could* be placed on the surface of Titan, sealed for heat and breathable air for plants and animals. The "could" is an open question, though. Habitats might have the tendency to melt the surface and sink or, actually, float away. Heat expands. Titan is so cold that any heat we add to the system rises up and would take a habitat with it as if it were a hot air balloon. Without understanding the consistency of the Titan surface, engineers cannot determine how to anchor the habitats. With flying the optimum mode of transportation, free-floating habitats might make sense if the levitation could be stabilized. Fun stuff.

The biggest drag for human dwellers on Titan, though, may be the lack of sunlight. The Sun is only 1 percent as bright to begin with compared with the rays on Earth, and the clouds cut that in half. Every day is a hazy day, no brighter than dusk on Earth. The biggest bummer is that you would barely be able to see lovely Saturn, which would appear more than ten times larger than a full moon on Earth and take up at least half the sky. You would need to don infrared glasses or otherwise ride in a stratospheric balloon to see it. Food, of course, needs to be grown under artificial lights.

Would you have alien neighbors? Life may have developed on Titan, although it would be like nothing we have seen before. Life doesn't necessarily need sun and water; it needs energy and a liquid medium. Titan has both. So, life would need to function without oxygen in this cold, liquid environment. Biologists have contemplated this and have developed a theory. They conceived a cell membrane that could be flexible at cryogenic temperatures made of carbon, hydrogen, and nitrogen, as opposed to the cell membranes we know, which are phospholipids: carbon, hydrogen, oxygen, and phosphorus. The hypothetical cell membrane is called an azotosome, which combines the French word for nitrogen, *azote,* and liposome.[13] But an azotosome is merely a cellular shell. How an azotosome-based creature would eat and reproduce remains unknown—life "not as we know it," as Isaac Asimov once wrote.

I have one more funky concept to share, introduced by Isaac Arthur, a physicist and YouTuber with a popular channel on futurism. Titan can serve as a massive heat sink, allowing industrial and computational activities to be far more efficient. Engines operate on the energy transfer between two reservoirs, one hot and one cold. Efficiency is measured as $E = 1 - T_C / T_H$, where temperature (T) is usually relayed in Kelvin. On Earth, a room or factory floor temperature is about 300K, and this is the cold reservoir, T_C. The engine might be operating with 400K heat reservoir, which would be T_H. The efficiency of this engine is $1 - 300K / 400K$, or 0.25. On Titan, the efficiency equation would look more like $1 - 100K / 400K$, or 0.75. This applies to supercomputers, which are known to create tremendous heat and which require enormous energy to cool. A supercomputer that is 25 percent efficient on Earth would be 75 percent efficient on Titan. Thus, Titan could serve as an ideal industrial base for humanity's manufacturing needs.

And now for the mind-blower: Isaac Arthur further speculates that, in the very distant future, should humans become virtual entities with brains uploaded into giant computers, we would need a place to house all those computers. The Earth won't do. The computers would generate more and more heat, warming Earth and driving down the efficiency. We need to house this operation on a place like Titan. Arthur calculates that Titan is large and cold enough to accommodate the computing power to contain trillions of human brains. Thus, all of humanity could one day wind up on Titan, protected even as the Sun expands in billions of years to consume Mercury, Venus, Earth, Mars, and Jupiter.[14]

Sweet Solitude of Uranus, Neptune, Pluto, and Beyond

Should you decide *not* to upload your brain into a supercomputer on Titan, there are still a few rewarding opportunities farther out into the Solar System.* Uranus, the next planet, has twenty-seven known moons, two of which, Titania and Oberon, could serve as a terrestrial base should the mining of Uranus prove feasible. Neptune has fourteen known moons, of which Triton is by far the largest and, with a possible liquid ocean under ice, a place to search for life. Pluto and the myriad trans-Neptunian objects, Kuiper belt, and Oort cloud objects have habitability potential by virtue of all the water ice they contain.

I should preface this by saying that there's no solid body out there that provides suitable gravity. The largest is Triton, with 0.08G. Underdog Pluto has only 0.06G. All ideas of settling on these objects or tethering to them with orbital rings reflect planetary chauvinism. The more logical approach to colonization would be to build orbiting cities with artificial gravity. Also, given the immense resources in the asteroid belt and the terrestrial opportunities around Jupiter and Saturn, it is hard to fathom the necessity of living in the outermost reaches of the Solar System.

*Note that uploading your brain is copy-paste, not cut-paste, so there would be two of you. (So you could live on Titan and explore the outer Solar System!)

The Sun is too distant there to power anything larger than a pocket calculator, and the entire prospect is heavily dependent on a fusion economy. In science fiction, people at this distance from the Sun are portrayed as the ultimate homesteaders, longing for supreme solitude. One other sci-fi trope is this notion of a dystopian, war-driven future. Here, those who control the outer Solar System control the inner Solar System, an extension of current terrestrial nations yearning for a Space Force. So, let's investigate.

Uranus and Its Shakespearean Moons

Uranus is more an ice giant than a gas giant, much smaller than Jupiter and Saturn. In books it looks bigger and is rarely drawn to scale. Uranus has the surface area of only sixteen Earths and an equatorial radius of 25,559 km, compared with 6,371 km on Earth. It is difficult to spot the equator, though, because the planet tilts 97 degrees and the poles essentially face the Sun. Each pole gets forty-two years of perpetual (dim) sunlight followed by forty-two years of darkness during Uranus's eighty-four-year journey around the Sun. Uranus also has a ring system, like Saturn. The planet's atmosphere is mostly hydrogen and helium but also has large amounts of methane and icy ammonia and water. Lots to work with there.

Someone long ago decided to name all of Uranus's moons after characters from the plays and poems of William Shakespeare and Alexander Pope, and who am I to complain. The moons are very small, and contemplating a settlement there may very well be much ado about nothing. Titania and Oberon are the largest, each about 1,500 kilometers across and sporting 0.04G. Both are ice rocks that could support a scientific or mining base in terms of providing water, but they seem otherwise unremarkable. The exciting object in this region is Uranus itself. With its relatively low mass, the escape velocity in the upper atmosphere is about the same as it is on Earth—and three times lower than it is on Jupiter. Combined with a surprisingly calm atmosphere and low wind speeds, this means we could swoop in and collect gas for export to the inner Solar System without exerting too much energy. Prime resources would be helium-3, should we have the fusion thing figured out, and nitrogen, in high demand on Mars and for filling those orbiting cities I've been pining about with an inert gas. So, hypothetically, a Uranian economy could emerge with bases, if needed, on Titania and Oberon. Titania and Oberon might even have subsurface liquid water harboring life. So little is known about these worlds.

Neptune and Triton

Neptune is much like Uranus in size, mass, rings, and composition, slightly smaller but denser. As an ice giant, it has roughly the same amount that Uranus has of hydrogen, helium, methane, and icy deposits of water and ammonia. One startling feature, however, is the speed of the winds in the atmosphere, which are the fastest recorded in the Solar System. Winds there have topped 2,100 km / h, making gas and ammonia mining too dangerous to attempt. Neptune does have a very interesting moon among its collection of fourteen, and that's Triton.

Triton is the seventh largest moon in the Solar System, slightly smaller than Europa and our Moon. The surface gravity is only 0.08G, and there's but a tenuous atmosphere. What makes Triton interesting, as you may have guessed, is a possible global subsurface ocean that could harbor life.[15] The moon erupts geysers of nitrogen from its surface, a sign of cryovolcanoes and radiogenic heating that could provide a food and energy source for life. In the planetary scheme of things, Triton is as strong as a contender for life as any moon around Jupiter or Saturn. Triton warrants a visit. A science base, yes. A colony, likely not.

One other interesting note is that Triton is the only moon in the Solar System to orbit its planet in the opposite direction. This is because Triton didn't form with Neptune. Rather, Neptune must have captured it from the Kuiper belt. Given its size, Triton may very well be considered a dwarf planet. Triton is larger and more massive than Pluto. This is more than just trivia; it implies that the Kuiper belt, which includes Pluto, likely contains an untold number of Triton-size objects, each with the possibility of hosting life in subsurface oceans. That plays into the panspermia concept, as well, the seeding of life from one planet to another. Life on Earth or on any planet or moon may have originated in a Kuiper belt objects that crashed onto its surface eons ago.

Fasten Your Kuiper Belt: Pluto, Eris, Sedna, and More

Everyone loves an underdog, which is perhaps why Pluto remains such a treasured Solar System object and a desired location for would-be settlers. But at the risk of stirring up some LUCA, there's nothing extraordinary about Pluto. Once designated a "regular" planet, it has since been demoted to dwarf planet status. One pragmatic reason is that there are likely hundreds of these kinds of objects in the Kuiper belt, a circumstellar disc of matter extending from the orbit of Neptune (30 AU) to approximately 50 AU.* And Pluto isn't even the most massive of the known objects. That distinction goes to Eris. Also, seven moons are larger than Pluto.

From a more theoretical point of view, however, the International Astronomical Union (IAU) has defined the concept of a

*An AU is an astronomical unit measured as the approximate distance between the Earth and Sun.

planet, in part, as an object that has "cleared the neighbourhood around its orbit."* Pluto's orbit is strongly influenced by Neptune, and it shares its orbit around the Sun with many other trans-Neptunian objects.

Pluto maintains distinction in our minds because it long held the position as the final and smallest planet. Also, the NASA New Horizons mission visited the Pluto region, sending back fabulous images of this icy world. So, we now know more about Pluto than about most planets and moons in the outer Solar System. But Pluto is nearly airless and has a surface gravity of 0.06, almost three times less than our own low-gravity Moon. A permanent settlement there seems impractical. Pluto's saving grace is that the dwarf planet is rich in volatiles such as hydrogen, oxygen, and nitrogen, all suitable for sustaining life. There's plenty of water ice. The problem is there's little of the heavy stuff, like iron and silicon, so there's less for building structures and supporting industry.

Tiny Pluto has five moons: Charon, Styx, Nix, Kerberos, and Hydra. This is where things get a little more interesting. Charon is the largest of these moons, more than half the diameter of Pluto, and some astronomers consider Pluto–Charon a binary dwarf planet system with irregularly shaped satellites left over from a past collision. What is curious about Pluto–Charon is the presence of macro-organic molecules called tholins, which give both objects a reddish-brown color in parts. Some scientists speculate that tholins are precursors of life, that they formed into more complex molecules, such as amino acids, naturally in the presence of water and other conditions on the early Earth. Many moons may have tholins, most notably Titan and Triton. According to data from

*From resolution 5A at the 26th General Assembly for the International Astronomical Union, 2006.

NASA's New Horizons, hydrocarbons in Pluto's thin atmosphere, bombarded by cosmic rays and solar ultraviolet radiation, are forming tholins, some of which are blowing over to Charon's north polar region, painting it red.[16] New Horizons has painted an entirely new view of the Pluto system, a place worth visiting. Long-term human occupation of Pluto, however, likely won't occur unless life is discovered there.

One very neat concept, though, is tethering Pluto to Charon to create an interplanetary highway. The two are only about 19,000 km apart, whereas the Earth and Moon are nearly 400,000 km apart. But what makes this theoretically possible is the fact that Pluto and Charon are tidally locked, so they both show the same face to each other. We see our Moon move across the sky. Those on Pluto would not see Charon budge, and vice versa. The tether system would need to be a little flexible, because the tidal lock is not perfect and the orbits shift ever so slightly. Nevertheless, this tether is doable even with modern materials. Folks could live in orbiting pods attached to the tether (providing artificial gravity at a healthy level) and ride the rails to work on Pluto or Charon, mining life-sustaining water, tholins, nitrogen, or ammonia for export to other deep-space colonies. This is sure fun to think about, however impractical given the probability of robotic mining replacing the need for any human to be way out there. (This is an irony of futurism: envisioning humans as farmers and miners, the two most primitive trades.)

Most of the larger objects in the Kuiper belt will prove to be scientifically interesting. Sadly, most of them are too far and dispersed to visit with any single probe. These so-called trans-Neptunian objects (TNO) of the Kuiper belt go by names such as Eris, Haumea, 2007 OR_{10}, Makemake, 50000 Quaoar, 90377 Sedna, 2002 MS_4, 90482 Orcus, and 120347 Salacia. In January

2019, *New Horizons* swung by 2014 MU_{69}, known by its nickname Ultima Thule. The visit was not because this object was deemed to be the most interesting—it's only 30 kilometers long, puny compared to the 2,000-km wide spherical Eris—but because it was relatively close to Pluto and *New Horizons* could get there with little fuel from its trajectory departing Pluto. We know so little about these objects that there's no use in speculating if, how, or when we would settle on them.

Comets: Catch a Wave and You're Sitting on Top of the World

Comets are ice-coated bodies that originate in the Kuiper belt and also from much farther out in the Oort cloud. They are characterized by their tails, or coma, which appear as they get closer to the Sun during their long and highly eccentric, elliptical orbits. The tail is a stunning visible display caused by the ice and other volatiles burning off with the heat of the Sun; the tail disappears as the comet returns to the outer Solar System. Among the more famous are Halley's Comet, which comes around Earth every seventy-four to seventy-nine years, and the brilliant Comet Hale–Bopp, discovered in 1995, which likely won't be seen again for another 2,300 years (unless we travel to it!). Visionary physicist Freeman Dyson viewed comets as perhaps the most habitable of all bodies in the Solar System. One thing is certain: if you land on one, you are in for a ride.

Humans could feasibly commandeer a comet as they would an asteroid. The gist is to land, hollow out the core, and insert a rotating habitat to create artificial gravity. Comets are more like frozen mud balls, not ice balls. The outer shell of ice would provide protection from radiation; the inner core of metals, minerals, and rocks would provide building materials. Most comets contain nearly every element we need for life. When shopping for a comet, you'd want a large one to provide enough space and stability for a colony of millions of people, which could be attained in a mountain-size comet tens of kilometers across.

To live in a comet, we would need fusion. Comets spend most of their days too far from the Sun for solar energy to be useful, and extracting hydrogen fuel from the ice to burn with oxygen would not create enough power to power a city. Using that hydrogen or deuterium in the muddy ice as a fusion fuel is another matter. Just a few kilograms would go a long way toward filling up the interior with artificial light. And here's the fun part: you could use that fusion fuel to power your comet as an interstellar spacecraft. Indeed, in terms of bang for the buck, a comet—with its kilometers-thick covering of ice and roomy interior—would offer the best protection from radiation and cosmic debris as you whiz across the galaxy at a velocity more than 10 percent of the speed of light. Swooping around the Sun for a gravitational kick and then firing your engines would provide a tremendous boost to speeds that could get you to the nearest stars in forty years.

Of course, this is a futuristic concept. But in terms of practicality, comet-based generational spaceships would be the most efficient way to reach the stars with hundreds of thousands of people in tow. We would more likely see humans commandeering comets than, say, living en masse on Pluto, because the latter would serve little purpose, however much easier it might be to accomplish.

Oort Cloud and the Great Beyond

Beyond the loosely defined boundary of the Kuiper belt and the edge of our Solar System lies the mystical Oort cloud, a region of interstellar space extending from about 0.8 to 3 light years from the Sun, or 10,000 to 50,000 AU, roughly halfway to the nearest star. The region is a theoretical construct; no direct observations have been made of it. Astronomers speculate that there are countless icy planetesimals there, loosely bound to the Solar System or, at best, more closely bound to our Sun gravitationally than to any other star. Some long-period comets and near-parabolic comets are thought to originate in the Oort cloud.

Modeling—that is, measuring the unknown—suggests the Oort cloud could have more than a hundred Earths' worth of stuff, or 100,000 times more than the asteroid belt.[17] Each solid body, though, is likely separated from the others by unfathomable distances, along the lines of Earth to Pluto, making for a rather lonely or secluded home. There could be rogue planets out here, too, which somehow have escaped their original solar systems. These also are called nomad or Steppenwolf planets, and there's no limit to the size range. A planet with Earth-like gravity could be out there waiting for us.

We won't reach the Oort cloud anytime soon. NASA's Voyager 1, our most distant probe, traveling close to 60,000 km / hour, is just now exiting the Solar System and will arrive at the inner edge of the Oort cloud in about 300 years—and take 30,000 years to fly through it.[18] So, how and why would we settle way out there?

The "how" would parallel our habitation on comets and asteroids, with an emphasis on fusion, because there's no solar or other suitable energy source available. The greatest challenge is communication. Distances are so great that it takes days to months to send a message from one Oort community to another, and that's implying there are satellite communication relays. The "why" is, for now, pure sci-fi. Perhaps, when feeling particularly down about the state of the world, you can envision a distant future dystopia in which evil saturates the Solar System, from Mercury clear out to Pluto and the Kuiper belt. Rogues have enslaved much of humanity and have destroyed Earth by bombarding it with the comets they command. Here, the Oort cloud is your refuge. You and your thousands of comrades could rather easily remain hidden in one of the billions or trillions of icy bodies. Assuming you don't reveal yourself by sending signals beyond your rock, the odds of your being found even by an advanced civilization is minuscule. The numbers and distances are on your side.

But if everyone is getting along in millennia to come, the Oort cloud could help support an intergalactic highway. Unlike shipping routes and terrestrial highways, intergalactic highways don't need midway stops for resting or refueling. In space, stopping only burns fuel and crushes the momentum you built up and desperately need for these multiyear passages from star to star at 10 percent light speed or faster. Oort cloud bodies—and note, most solar systems likely have a similar clouds of bodies—could serve as cosmic lighthouses. They could be beacons for navigation but also use that light to propel ships.[19] (Later in this chapter I discuss solar sails, in which photons from the Sun could puff the sails of a large spacecraft and propel it to a discernable fraction of light speed.) The Sun's light is more dispersed and thus imparts a weaker thrust on the sails farther out, of course. So, beacons in the Oort cloud could beam powerful, focused lasers in various directions as a wind for spacecraft to catch, either to another solar system or to the inner planets of our Solar System. Depending to the scale of intergalactic trade, Oort cloud residents could earn a good living controlling the trade winds.

Whom would we be trading with? Likely ourselves, for there appears to be no sign of humanlike intelligence in the local galaxy in terms of radio signaling or similar manipulation of the electromagnetic spectrum. And should other intelligent life exist, the chance of these creatures being keen on capitalism is rather slim.

In one possible scenario, centuries before we establish trading networks, we may set out to new lands in what are called generation starships or interstellar arks, starships that travel at sub-light speed. The concept here can be grasped easily enough: these would be massive, self-sufficient spacecraft on a journey to another star that would take hundreds or thousands of years. Thus, generations of occupants would live and die on the spacecraft, unless we had figured out the immortality thing by that point.

These ships need to be fairly large. Ships in the current fleet of ocean cruise lines on Earth are more than three football fields long, yet carry only 5,000 to 6,000 passengers, which may seem like hell if you are on one but, in the cosmic scheme of things, is barely enough people to support a new space colony. The interstellar arks would need to be built in space; and if we go in that direction, the first generation of ships might be built from raw materials taken from the Moon or asteroids. Recall, asteroids or comets could be hollowed out to form generation starships, too. This brings up another interesting concept: a "primitive" ark that sets sail for Alpha Centauri in the year 2200 might be overtaken by an advanced ark traveling much faster, leaving in 2250. Pioneering passengers on the first ship might arrive at their destination utterly surprised that humans have already been there for one hundred years. But space is big, and hopefully they won't fight over the same solar system.

Need for Speed

Lingering ocean cruises may be nice if you're not in a hurry. I can think of no benefit of taking your time in space, though. There's no scenery to speak of, and you're sailing through a storm of deadly radiation. The faster you get to your destination the better. Indeed, the relative lack of speed is the limiting factor in our exploration of deep space. Even if we build spacecraft to protect against zero gravity, radiation, and other cosmic debris, how can we effectively link humanity across the Solar System, let alone the stars, when it currently takes ten years to get from Earth to Pluto? And that's going one way. Sailors may have spent years on the sea, but not decades.

As discussed in Chapter 3, rocket launches require sheer thrust with chemical or nuclear fuels to escape from Earth's gravity well. By the next century, when humans are poised to explore deep space, rockets actually may be obsolete. Skyhooks and orbital rings are far more efficient in lifting humans into space, where we can board waiting spacecraft. Then, once we are in space, we have more fuel options. Some of these options could propel us to a respectable percentage of light speed.

Ion and Plasma Propulsion

Ion propulsion is a marriage of the tortoise and the hare, a system to propel a spacecraft, via a series of atomic-size nudges, to immense speeds. Both NASA and the Japan Aerospace Exploration Agency (JAXA) have used ion propulsion successfully on space missions to asteroids. At the heart of this technology is the action–reaction concept: pushing a positively charged atom, or ion, of gas out the back end to propel the spacecraft forward by the same energy. Because there's no air resistance in space, the spacecraft will move faster and faster with each nudge.

Chemical fuels expel hot gas through a nozzle at about 5 km / sec. With a lot of this fuel, we can generate the thrust to lift the rocket. Then the fuel is gone, and you're left at that final speed attained by the launch. Ion thrusters use xenon gas. When bombarded with electrons, the xenon atoms lose an electron and become positively charged ions, which can be accelerated in an electric field and ejected at 40 km / sec. Ion by ion, out they go; the thrust is tiny, about 0.5 newtons, or the force to lift a sheet of paper. In space, though, the thrust adds up. The NASA *Dawn* spacecraft that visited the Ceres and Vesta asteroids used ion propulsion (once it was placed in space with a traditional rocket) to accelerate from 0 to 100 km / hour in four days, not exactly highway speed but ideal for the fine maneuvering required on this mission. Given a few weeks, a spacecraft with ion thrusters could reach 320,000 km / hour. That speed could reduce the current travel time to Mars to just a few months. We could reach Pluto in about five years instead of the nearly ten years it took *New Horizons.*

The current, working ion thruster is fine for light spacecraft but not for one with a heavy load—there would be too much mass to push. NASA has successfully demonstrated a more efficient ion propulsion system, albeit in laboratories on Earth. It's called an X3 or Hall thruster, and it can generate 5 newtons of thrust, which is tenfold higher than the xenon thruster used on *Dawn.* This means that an X3 propulsion engine could propel a reasonably hefty load to Mars, the kind of serious cargo needed to establish a science base or settlement.

Far more deeply in the experimental realm is the Variable Specific Impulse Magnetoplasma Rocket, or VASIMR, under development by the Ad Astra Rocket Company and led by former NASA astronaut Franklin Chang Díaz. Whereas the current ion propulsion systems use solar panels to create electrons to bombard xenon gas, VASIMR uses radio waves to "boil" electrons out of argon gas to create a plasma of ions. Engineers at Ad Astra have calculated that a spacecraft with a small nuclear reactor power source could generate an energetic ion plasma propellant to reach Mars is only thirty-nine days, compared to about 200 days with chemical fuels.[20]

Alas, in the fantasy realm, it now seems, is the EmDrive, a hypothesized type of propellant-free drive that breaks various laws of physics yet had folks testing it nonetheless for more than a decade. The concept was that microwaves collected in a cone-like device could bounce around and create minuscule thrust. Experimenters, including some at NASA, thought they saw this effect. If it could work, you could power spacecraft with ambient microwave space radiation. Clearly this would be ideal for interstellar travel; with no fuel you could reach enormous speeds. But a team of engineers in Germany have found that the detected thrust in earthbound laboratories was from power cables in the engine chamber interacting with the Earth's magnetic field.[21]

Solar Sails

Solar sails catch the solar wind or, more properly, photon pressure from sunlight. JAXA was the first to prove this technology in interplanetary space, using solar sails to send IKAROS (Interplanetary Kite-craft Accelerated by Radiation of the Sun) to Venus in 2010. The sail was 14 meters by 14 meters and just a few microns thick, and this propelled the 315-kilogram craft to a maximum velocity of 1,440 km/hour, or 0.4 km/sec. While that's much slower than what ion thrusters can muster, the potential is still great. Note that IKAROS was sailing head into the "wind." In 2019, with money raised through crowdfunding, the Planetary Society successfully launched and deployed a solar sail to demonstrate the ability to move spacecraft around the Solar System with no fuel.

In theory, a solar sail whipping around close to the Sun, within Mercury's orbit, could catch enough breeze to send it sailing

400 km/sec, or 0.001 light speed.[22] That's fast enough to get to Pluto in only about two years. A spacecraft with sails can go even faster if we push it with laser beams. The Breakthrough Starshot initiative is a project to send a thousand tiny space probes to the Alpha Centauri star system, four light-years away. Powerful lasers on Earth would push these probes, just a few centimeters wide, and accelerate them to 15 to 20 percent light speed. One catch with this otherwise brilliant scheme is slowing the probes once they reach Alpha Centauri. Parachutes don't work in space.

Possible and Improbable Propulsion

Humans can dream, and that's essentially what the folks working for NASA's Breakthrough Propulsion Physics (BPP) Project did from 1996 until 2002, when the project was canceled. There was some math and physics sprinkled in there, but mostly it was dreaming. One of the concepts investigated was warp drive, à la *Star Trek*. This isn't quite faster-than-light travel, as popularly conceived. Rather, a warp drive warps space—squishes it together—allowing you to traverse great distances by skipping across the crest of waves instead of up and down each ripple.

You didn't miss the press release: warp drive didn't pan out. Warp drive and its funky twin, wormholes, are concepts said to be feasible because they don't break known laws of physics. The energy required to distort space, however, is far beyond anything we can generate until, perhaps, we learn to harness the power of black holes.

Antimatter fuel is in the realm of the possible. We can make antimatter today in particle accelerators, although it is difficult to store for a long time. And the quantities we produce are literally just a few antiprotons, less than a billionth of a gram. Antimatter refers to otherwise identical particles that have opposite charges. An antielectron, called a positron, has a positive charge, the opposite of an electron's negative charge. An antiproton has a negative charge. They are highly unstable, and when antimatter meets ordinary matter, the particles annihilate. No ash—a pure transfer of matter into energy, as described by the equation $E=mc^2$. Chemical energy is approximately 1 percent efficient; lots of ash left over. Nuclear energy is about 10 percent efficient. Matter–antimatter annihilation is 100 percent efficient. All this is to say that if we ever harness the power of antimatter—and this isn't entirely infeasible—we would have a fuel to propel us at upward of 40 percent light speed. At this point, we'd need to worry about moving so fast, because plowing through cosmic debris at that speed will eat away at the hull of the ship. If you think bugs on the car windshield are bad . . .

Closer to reality are fission and fusion engines. Nuclear fuel is used in space now. NASA loaded up Voyagers 1 and 2 with radioisotope thermoelectric generators (RTGs), and those probes are now exiting the Solar System. Heat from decaying plutonium generates electricity. RTGs are used on several probes, such as the Mars *Curiosity* rover. Nuclear *fission* in space, though, has been a challenge. NASA's Nuclear Engine for Rocket Vehicle Application (NERVA) program ran for two decades, in the 1950s and 1960s. This fission-powered rocket was to take humans to Mars by the 1980s but proved too costly to develop further. The program was reborn in the late 1980s as Project Timberwind, funded by the Strategic Defense Initiative ("Star Wars"). The problem then and now is the safety of nuclear fuel, particularly in lifting the spacecraft off the Earth. If the rocket explodes, then a large swath of terrain would be dusted with toxic nuclear fuel—a human nightmare and indeed a political one, should the rocket of one country explode over another. The nuclear engine technology is well advanced, though. So, fission rockets could be feasible if launched from, say, the lifeless Moon with nuclear fuel created on the Moon, bypassing the safety concerns.

Should we master fusion, there's an even greater potential to launch massive spacecraft from the Moon powered by local helium-3 deposits. Compared with chemical fuels, nuclear fuels not only provide more power but also burn more efficiently. This means we could use nuclear fuels to go faster and longer, with a much smaller fuel-to-cargo ratio. Pulsed fusion, in which just a tiny amount of fusion fuel is used at a time to create a series of pushes, could propel a spacecraft to 10 percent of light speed.

My prediction: Powered by sails and ion propulsion in large ships assembled in space, after scientific bases have been established on the Moon and Mars, humans visit the skies of Venus and moons of Jupiter by the end of the twenty-first century; life on at least one of the outer Solar System moons is discovered by a robotic mission by the end of the twenty-first century; in blisteringly fast nuclear-powered spacecraft, humans venture to Titan by the early twenty-second century; science and technology have advanced by the end of the twenty-second century to allow for human scientific exploration throughout the Solar System, yet there remains no need for commercial activity or residency beyond Uranus, as this ice giant is sufficient to supply all necessary resources to the inner Solar System; the first humans leave for the closest, habitable solar system by the end of the twenty-third century; millennia will pass before dots are connected and interstellar travel and commerce becomes the norm.

Epilogue

Welcome Home

The Earth will be our home for millennia to come. No concept introduced in this book implies that humanity will be leaving Earth anytime soon. Yes, the Earth has problems. But leaving Earth to escape problems or an assumed imminent danger is both impractical and foolish. I can see no realistic scenario in which, for example, living on Mars would be better than living on Earth. Aside from utter disintegration of the mother planet caused by a moon-size collision or a Vogon construction fleet paving an intergalactic superhighway, as was the plot in *The Hitchhiker's Guide to the Galaxy,* nothing could make Earth less hospitable than Mars. Consider nuclear war or asteroid strike, which have essentially the same result. The remaining percent of a percent of the human race would have an easier time living underground and surviving for a few years on canned beans lifted from the rubble of a supermarket than starting a new life for humankind on frigid, sterile, nearly airless Mars. And remember, living on any other planet or moon in this great Solar System of ours is actually more challenging than living on Mars.

Human space exploration is not a plan B for Earth. Rather, our activities in space make living on Earth better. Weather satellites alert us to the direction and magnitude of storms days in advance; communication satellites drive the global economy; Earth–Sun observatories illuminate trends in atmospheric pollution and the rise of greenhouse gases; orbiting space-science satellites the likes of the Hubble Space Telescope and the Wilkinson Microwave Anisotropy Probe answer profound questions about the age and composition of the universe and provide us with a sense of wonder. Space technology and exploration is far from futuristic escapism. It is the here and now.

I see our presence on the Moon, Mars, and beyond as a means of helping guarantee our health on Earth, securing what President Franklin D. Roosevelt called "freedom from want" and the right to an adequate standard of living. This includes access to clean water and food and the elimination of the disparities that permit the degradation of land, the breaking of backs, the crushing of hands, the corruption of lungs, and the wasting of minds. By entering space and diversifying our global economy, we open new possibilities for energy and resource extraction, while at the same time inspiring younger generations to look outward and to reverse the relapse into petty tribalism that is now affecting nearly every nation.

Our exploration of the Solar System will change Earth in ways we cannot predict. But it would be unprecedented and run counter to the chronicles of human history if we were to find—fifty, one hundred, two hundred years from now—that the Earth was worse off for venturing into space. Some fear technological developments, it's true. Yet technology can protect the sanctity of primitivism. Space resources, for example, can reduce the dependency on coal, oil, timber, and precious minerals on Earth, allowing the hunting-and-gathering cultures of the Amazon and Southeast Asia to live their natural lives free from the encroachment of ruthless commercial interests. Technology enables living off the grid because the tools of such an existence—solar panels, water purification, wireless communication, and Internet-based learning—are space age. Technology can ameliorate terrorism, too, as the diversification of resource extraction minimizes conflicts concerning terrestrial claims to land and water.

I would like to see a future in which there are no losers in the exploration game, no native populations wiped out by conquistadors, no populations exploited for cheap labor. European wealth and, to a great extent, US wealth, is infamously built on this history of exploitation. In contrast to Asian countries, which largely transplanted their populations to new lands and kept the wealth there, European and American colonies removed that wealth (minerals, timber, etc.) back to their homelands at the expense of the native populations. I don't mean to wax political with talk of exploitation of the worker. Indeed, the concepts presented in this book are aligned with capitalism, not communism. The point is that in space, on lifeless bodies such as the asteroids, the Moon, or Mars, the extraction of resources affects no alien population. These resources could be brought to Earth to place humanity on the track toward a post-scarcity dawn where there is no longer fear of want.

Tapping into the unlimited natural resources of space would allow the human population to expand. It is with a sincere sense of humanity that I say that enabling billions or even trillions of beings to experience a dignified human life is a fine goal; and frankly I have trouble understanding the counterargument that we need to halt or reduce the population level as a result of dwindling resources. Returning to Earth in the twenty-third century, after a long journey through the Solar System in which I visited the multitude of residential settlements on Mars and dotted throughout the Earth–Moon system in orbiting cities, as well as the robotic work settlements of the asteroid belt and slightly beyond, I would expect to see a global population in the tens of billions living in the efficiencies that space resources can bring in terms of energy and materials.

Humans don't *have* to go into space. And although we are bound to venture back to the Moon and then to Mars within the next few decades, there's a solid chance that we cannot stay there, at least not in this century, because the physical challenges are too great and the economic rewards too small. Yet at some point—it could be the next decade or the next century—it will simply, somehow, just make sense to be in space. Space will be a natural extension of humanity, as was our bridging water and then air. And when that era arrives, all of humanity may prosper, and *Homo sapiens* will take the first bold leap toward the evolution of *Homo futuris.*

Notes

Introduction

1. Thomas Heppenheimer, *The Space Shuttle Decision: NASA's Search for a Reusable Space Vehicle* (Washington, DC: NASA History Office SP-4221, 1999), 146, https://ntrs.nasa.gov/archive/nasa/casi.ntrs.nasa .gov/19990056590.pdf.
2. John M. Logsdon, "Ten Presidents and NASA," *50th Magazine—50 Years of Exploration and Discovery,* NASA (2008), https://www.nasa.gov/50th/50th_magazine/10presidents.html; FY 2018 Budget Request, NASA, https://www.nasa.gov/content/fy -2018-budget-request.
3. John M. Logsdon, *After Apollo?: Richard Nixon and the American Space Program* (London: Palgrave Macmillan, 2015).
4. Logsdon, *After Apollo?*
5. Zuoyue Wang, *In Sputnik's Shadow: The President's Science Advisory Committee and Cold War America* (New Brunswick, NJ: Rutgers University Press, 2009), 222.
6. Roald Z. Sagdeev, *The Making of a Soviet Scientist: My Adventures in Nuclear Fusion and Space from Stalin to Star Wars* (Hoboken, NJ: Wiley, 1994).
7. Heppenheimer, *Space Shuttle Decision,* 115.
8. William Sims Bainbridge, "The Impact of Space Exploration on Public Opinions, Attitudes, and Beliefs," in *Historical Studies in the Societal Impact of Spaceflight,* ed. Steven J. Dick (NASA, 2015).

9. Richard Nixon, "Statement about the Future of the United States Space Program," March 7, 1970; online by Gerhard Peters and John T. Woolley, American Presidency Project, http://www.presidency.ucsb.edu/ws/?pid=2903.
10. "Space Station: Staff Paper Prepared for the President's Commission to Study Capital Budgeting," Clinton White House archives, June 19, 1998, https://clintonwhitehouse5.archives.gov/pcscb/rmo_nasa.html.
11. Neil deGrasse Tyson, "Paths to Discovery," in *The Columbia History of the Twentieth Century,* ed. Richard W. Bulliet (New York: Columbia University Press, 1998), 461–482.
12. Neta C. Crawford, "US Budgetary Costs of Wars through 2016: $4.79 Trillion and Counting," White Paper (Providence, RI: Brown University, 2016).

1. Living on Earth

1. Claude Lafleur, "Costs of US Piloted Programs," *Space Review,* March 8, 2010, http://www.thespacereview.com/article/1579/1.
2. United Nations Department of Economic and Social Affairs, Population Division, "World Population Prospects: The 2017 Revision, Key Findings and Advance Tables," 2017, https://esa.un.org/unpd/wpp/Publications/Files/WPP2017_KeyFindings.pdf.
3. Patrick Gerland et al., "World Population Stabilization Unlikely This Century," *Science,* 346 (2014): 234–237, doi:10.1126/science.1257469.
4. Dana Gunders, "Wasted: How America Is Losing Up to 40 Percent of Its Food from Farm to Fork to Landfill," NRDC Issue Paper, August 2012, https://www.nrdc.org/sites/default/files/wasted-food-IP.pdf; Lawrence Livermore National Laboratory, "Americans Continue to Use More Renewable Energy Sources," press release, July 18, 2013, https://www.llnl.gov/news/americans-continue-use-more-renewable-energy-sources.
5. Peter A. Curreri and Michael K. Detweiler, "A Contemporary Analysis of the O'Neill-Glaser Model for Space-Based Solar Power and Habitat Construction," NASA Technical Reports Server, 2011, doi:10.1061/41096(366)117.

6. Barbara Tuchman, *A Distant Mirror: The Calamitous 14th Century* (New York: Alfred A. Knopf, 1978).
7. Karen Meech et al., "A Brief Visit from a Red and Extremely Elongated Interstellar Asteroid," *Nature* 552 (2017): 378–381, doi:10.1038/nature25020.
8. Philip Plait, private communication, April 8, 2018.
9. Food and Agriculture Organization of the United Nations, "The State of Food Security and Nutrition in the World" (FAO, 2017), http://www.fao.org/3/a-I7695e.pdf.
10. Patrick T. Brown and Ken Caldeira, "Greater Future Global Warming Inferred from Earth's Recent Energy Budget," *Nature* 552 (2017): 45–50, doi:10.1038/nature24672.
11. World Bank Group, "Turn Down the Heat: Confronting the New Climate Normal" (World Bank, 2014), http://documents.worldbank.org/curated/en/317301468242098870/Main-report.
12. Brian Thomas et al., "Gamma-Ray Bursts and the Earth: Exploration of Atmospheric, Biological, Climatic and Biogeochemical Effects," *Astrophysical Journal* 634 (2005): 509–533, doi:10.1086/496914.
13. Adrian L. Melott and Brian C. Thomas, "Late Ordovician Geographic Patterns of Extinction Compared with Simulations of Astrophysical Ionizing Radiation Damage," *Paleobiology* 35 (2009): 311–320, arXiv:0809.0899.
14. Julia Zorthian, "Stephen Hawking Says Humans Have 100 Years to Move to Another Planet," *Time,* May 4, 2017, http://time.com/4767595/enums; Justin Worland, "Stephen Hawking Gives Humans a Deadline for Finding a New Planet," *Time,* November 17, 2016, http://time.com/4575054/stephen-hawking-humans-new-planet.
15. Peter Gillman and Leni Gillman, *The Wildest Dream: The Biography of George Mallory* (Seattle: Mountaineers Books, 2001), 222.
16. Colin Summerhayes and Peter Beeching, "Hitler's Antarctic Base: The Myth and the Reality," *Polar Record* 43 (2007): 1–21, doi:10.1017/S003224740600578X.
17. United States Department of State, "The Antarctic Treaty," https://www.state.gov/documents/organization/81421.pdf.
18. United Nations Office for Outer Space Affairs, "Treaty on Principles Governing the Activities of States in the Exploration and Use

of Outer Space, including the Moon and Other Celestial Bodies," http://www.unoosa.org/pdf/gares/ARES_21_2222E.pdf.

19. Mahlon C. Kennicutt et al., "Polar Research: Six Priorities for Antarctic Science," *Nature* 512 (2014): 23–25, doi:10.1038/512023a.
20. Christopher Wanjek, "Food at Work: Workplace Solutions for Malnutrition, Obesity and Chronic Diseases" (Geneva: ILO, 2005), 138–142.
21. German Aerospace Center (DLR), "Rich Harvest in the Antarctic EDEN ISS Greenhouse—Tomatoes and Cucumbers in the Polar Night," June 25, 2018, https://www.dlr.de/dlr/en/desktopdefault.aspx/tabid-10081/151_read-28538/.
22. Martin Gorst (Director), *Big, Bigger, Biggest: Submarine,* Documentary, Windfall Films (National Geographic International, 2009).
23. Gorst, *Big, Bigger, Biggest.*
24. Thomas Limero et al., "Preparation of the NASA Air Quality Monitor for a U.S. Navy Submarine Sea Trial," Conference Paper, Submarine Air Monitoring and Air Purification Symposium, Uncasville, Connecticut, November 13–16, 2017.
25. Zoltan Barany, *Democratic Breakdown and the Decline of the Russian Military* (Princeton, NJ: Princeton University Press, 2007), 33–34.
26. Ned Quinn, "NASA and U.S. Submarine Force: Benchmarking Safety," *Undersea Warfare: The Official Magazine of the U.S. Submarine Force* 7 (fall 2005).
27. Katie Shobe et al., "Psychological, Physiological, and Medical Impact of the Submarine Environment on Submariners with Application to Virginia Class Submarines," Naval Submarine Medical Research Laboratory Technical Report #TR-1229 (2003), https://www.researchgate.net/publication/256687350.
28. "How to Survive the Long Trip to Mars? Ask a Submariner," Associated Press via *Washington Post,* October 5, 2015, https://www.washingtonpost.com/lifestyle/kidspost/how-to-survive-the-long-trip-to-mars-ask-a-submariner/2015/10/05/9d05bf82-6b82-11e5-9bfe-e59f5e244f92_story.html.
29. Peter Rejcek, "Passing of a Legend: Death of Capt. Pieter J. Lenie at Age 91 Marks the End of an Era in Antarctica," *Antarctic Sun,* April 20, 2015, https://antarcticsun.usap.gov/features/contenthandler.cfm?id=4150.

30. Vadim Gushin et al., "Content Analysis of the Crew Communication with External Communicants under Prolonged Isolation," *Aviation, Space, and Environmental Medicine* 12 (1997): 1093–1098.
31. Robin Marantz Henig, "Cabin Fever in Space," *Washington Post,* August 4, 1987.
32. Ian Mundell, "Stop the Rocket, I Want to Get Off," *New Scientist,* April 17, 1993, pp. 34–36, https://www.newscientist.com/article/mg13818693-700.
33. "HI-SEAS Hawaii Space Exploration Analog and Simulation," University of Hawai'i at Mānoa media kit, September 2017.
34. Leslie Mullen, "A Taste of Mars: Hi-Seas Mission Now in Its Final Days," *Astrobiology Magazine,* August 8, 2013, https://www.astrobio.net/moon-to-mars/a-taste-of-mars-hi-seas-mission-now-in-its-final-days.
35. "Second HI-SEAS Mars Space Analog Study Begins," University of Hawai'i at Mānoa press release, March 28, 2014, https://manoa.hawaii.edu/news/article.php?aId=6399.
36. Julielynn Y. Wong and Andreas C. Pfahnl, "3D Printed Surgical Instruments Evaluated by a Simulated Crew of a Mars Mission," *Aerospace Medicine and Human Performance* 87 (2016): 806–810, doi:10.3357/AMHP.4281.2016.
37. Allison Anderson et al., "Autonomous, Computer-Based Behavioral Health Countermeasure Evaluation at HI-SEAS Mars Analog," *Aerospace Medicine and Human Performance* 87 (2016): 912–920, doi:http://10.3357/AMHP.4676.2016.
38. Michael Tabb, "This Team Is Simulating a Mission to Mars to Understand the High Emotional Cost of Living There," *Quartz,* March 10, 2016, https://qz.com/635323/this-team-is-faking-a-mission-to-mars-to-understand-the-high-emotional-cost-of-living-there.
39. Ilya Arkhipov, "Russia Picks Space-Pod Team for 520-Day Moscow 'Voyage' to Mars," *Bloomberg Businessweek,* May 18, 2010, https://www.bloomberg.com/news/articles/2010-05-18/russia-picks-moscow-space-pod-team-for-520-day-simulated-voyage-to-mars.
40. Yue Wang et al., "During the Long Way to Mars: Effects of 520 Days of Confinement (Mars500) on the Assessment of Affective Stimuli and Stage Alteration in Mood and Plasma Hormone Levels," *PLoS ONE* 9 (2014), doi:10.1371/journal.pone.0087087.

41. Mathias Basner et al., "Psychological and Behavioral Changes during Confinement in a 520-Day Simulated Interplanetary Mission to Mars," *PLoS ONE* 9 (2014), doi:10.1371/journal.pone.0093298.
42. Peter Suedfeld, "Historical Space Psychology: Early Terrestrial Explorations as Mars Analogues," *Planetary and Space Science* 58 (2010): 639–645.
43. Jack W. Stuster, "Bold Endeavors: Behavioral Lessons from Polar and Space Exploration," *Gravitational and Space Biology Bulletin* 13 (2000): 49–57.
44. Stuster, "Bold Endeavors," 53.
45. Stuster, "Bold Endeavors," 53.
46. Jane Poynter, *The Human Experiment: Two Years and Twenty Minutes Inside Biosphere 2* (New York: Basic Books, 2006).
47. Rebecca Reider, *Dreaming the Biosphere: The Theater of All Possibilities* (Albuquerque: University of New Mexico Press, 2009).

2. Checkup before Countdown

1. Agence France Presse, "Astronaut Vision May Be Impaired by Spinal Fluid Changes," November 28, 2016, https://www.yahoo.com/news/astronaut-vision-may-impaired-spinal-fluid-changes-study-191419024.html.
2. Donna R. Roberts et al., "Effects of Spaceflight on Astronaut Brain Structure as Indicated on MRI," *New England Journal of Medicine* 377 (2017): 1746–1753, doi:10.1056/NEJMoa1705129.
3. Gil Knier, "Home Sweet Home," NASA MSFC, May 25, 2001, https://web.archive.org/web/20060929044226/http://liftoff.msfc.nasa.gov/news/2001/news-homehome.asp.
4. Dai Shiba et al., "Development of New Experimental Platform 'MARS'—Multiple Artificial-gravity Research System—to Elucidate the Impacts of Micro / Partial Gravity on Mice," *Scientific Reports* 7 (2017), doi:10.1038/s41598-017-10998-4.
5. Michael J. Carlowicz and Ramon E. Lopez, *Storms from the Sun* (Washington, DC: Joseph Henry Press, 2002), 144.
6. Tony Phillips, "Sickening Solar Flares," NASA, November 8, 2005, https://www.nasa.gov/mission_pages/stereo/news/stereo_astronauts.html.

7. Lawrence Townsend et al., "Extreme Solar Event of AD775: Potential Radiation Exposure to Crews in Deep Space," *Acta Astronautica* 123 (2016): 116–120, doi:10.1016/j.actaastro.2016.03.002.
8. Christer Fuglesang et al., "Phosphenes in Low Earth Orbit: Survey Responses from 59 Astronauts," *Aviation, Space, and Environmental Medicine* 77 (2016): 449–452.
9. Eugene N. Parker, "Shielding Space Travelers," *Scientific American* 294 (2006): 40–47, doi:10.1038/scientificamerican0306-40.
10. Vipan K. Parihar et al., "Cosmic Radiation Exposure and Persistent Cognitive Dysfunction," *Scientific Reports* 6 (2016), doi:10.1038/srep34774.
11. Christopher Wanjek, "On a Long Trip to Mars, Cosmic Radiation May Damage Astronauts' Brains," *Live Science,* October 11, 2016, https://www.livescience.com/56449-cosmic-radiation-may-damage-brains.html.
12. Catherine M. Davis et al., "Individual Differences in Attentional Deficits and Dopaminergic Protein Levels Following Exposure to Proton Radiation," *Radiation Research* 181 (2014): 258–271, doi:10.1667/RR13359.1; Melissa M. Hadley et al., "Exposure to Mission-Relevant Doses of 1 GeV / n^{48}Ti Particles Impairs Attentional Set-Shifting Performance in Retired Breeder Rats," *Radiation Research* 185 (2016): 13–19, doi:10.1667/RR14086.1.
13. Jonathan D. Cherry et al., "Galactic Cosmic Radiation Leads to Cognitive Impairment and Increased Ab Plaque Accumulation in a Mouse Model of Alzheimer's Disease," *PLoS ONE* 7 (2012): e53275, doi:10.1371/journal.pone.0053275.
14. Christina A. Meyers and Paul D. Brown, "Role and Relevance of Neurocognitive Assessment in Clinical Trials of Patients with CNS Tumors," *Journal of Clinical Oncology* 24 (2006): 1305–1309, doi:10.1200/JCO.2005.04.6086.
15. Francis A. Cucinotta and Eliedonna Cacao, "Non-Targeted Effects Models Predict Significantly Higher Mars Mission Cancer Risk than Targeted Effects Models," *Scientific Reports* 7 (2017), doi:10.1038/s41598-017-02087-3.
16. IOM (Institute of Medicine), *Health Standards for Long Duration and Exploration Spaceflight: Ethics Principles, Responsibilities, and Decision Framework* (Washington, DC: National Academies Press, 2014), 1–4.

17. Amanda L. Tiano et al., "Boron Nitride Nanotube: Synthesis and Applications," Proceedings, SPIE, vol. 9060, Nanosensors, Biosensors, and Info-Tech Sensors and Systems, April 16, 2014, doi:10.1117/12.2045396.
18. Theo Vos et al., "Global, Regional, and National Incidence, Prevalence, and Years Lived with Disability for 310 Diseases and Injuries, 1990–2015: A Systematic Analysis for the Global Burden of Disease Study 2015," *Lancet* 388 (2016): 1545–1602, doi:10.1016/S0140-6736(16)31678-6.
19. Chad G. Ball et al., "Prophylactic Surgery Prior to Extended-Duration Space Flight: Is the Benefit Worth the Risk?" *Canadian Journal of Surgery* 55 (2012): 125–131, doi:10.1503/cjs.024610.
20. Francine E. Garrett-Bakelman et al., "The NASA Twins Study: A Multidimensional Analysis of a Year-Long Human Spaceflight," *Science* 364 (6436): eaau8650, doi:10.1126/science.aau8650.
21. Scott Kelly, *Endurance: A Year in Space, A Lifetime of Discovery* (New York: Knopf, 2017).
22. NASA-NIH, "Memorandum of Understanding between the National Institutes of Health and the National Aeronautics and Space Administration for Cooperation in Space-Related Health Research," September 12, 2007, https://www.niams.nih.gov/about/partnerships/nih-nasa/mou.
23. Wendy Fitzgerald et al., "Immune Suppression of Human Lymphoid Tissues and Cells in Rotating Suspension Culture and Onboard the International Space Station," *In Vitro Cellular & Developmental Biology—Animal* 45 (2009): 622–632, doi:10.1007/s11626-009-9225-2.
24. Sara R. Zwart et al., "Vitamin K Status in Spaceflight and Ground-Based Models of Spaceflight," *Journal of Bone and Mineral Research* 26 (2011): 948–954, doi:10.1002/jbmr.289.
25. Sarah L. Castro-Wallace et al., "Nanopore DNA Sequencing and Genome Assembly on the International Space Station," *Scientific Reports* 7 (2017), doi:10.1038/s41598-017-18364-0.
26. Kate Rubins, "An Afternoon with NASA Astronaut Kate Rubins," live event at National Institutes of Health, Bethesda, Maryland, April 25, 2017.

3. Living in Orbit

1. Roald Z. Sagdeev, *The Making of a Soviet Scientist: My Adventures in Nuclear Fusion and Space from Stalin to Star Wars* (Hoboken, NJ: John Wiley & Sons, 1994), 5.
2. Phillip F. Schewe, *Maverick Genius: The Pioneering Odyssey of Freeman Dyson* (New York: St. Martin's Griffin, 2014).
3. Ranga P. Dias and Isaac F. Silvera, "Observation of the Wigner-Huntington Transition to Metallic Hydrogen," *Science* 355 (2017): 715–718, doi:10.1126/science.aal1579.
4. Isaac Silvera, private communication, March 20, 2018.
5. Roger A. Pielke Jr., "The Rise and Fall of the Space Shuttle," *American Scientist* 96 (2008): 432–433.
6. Traci Watson, "NASA Administrator Says Space Shuttle Was a Mistake," *USA TODAY,* September 27, 2005, https://usatoday30.usatoday.com /tech/science/space/2005-09-27-nasa-griffin-interview_x.htm.
7. Michael Griffin, *Leadership in Space: Selected Speeches of NASA Administrator Michael Griffin, May 2005–October 2008* (Washington, DC: NASA, 2008), 328.
8. Sagdeev, *Making,* 213–214.
9. Andrew Chaikin, "Is SpaceX Changing the Rocket Equation?" *Air & Space Magazine,* January 2012, https://www.airspacemag.com /space/is-spacex-changing-the-rocket-equation-132285884/.
10. Christian Davenport, *The Space Barons: Elon Musk, Jeff Bezos, and the Quest to Colonize the Cosmos* (New York: PublicAffairs, 2018)
11. Davenport, *Space Barons.*
12. Cristina T. Chaplain, "The Air Force's Evolved Expendable Launch Vehicle Competitive Procurement," letter to US Congress, GAO-14-377R Space Launch Competition, March 4, 2014, https://www .gao.gov/assets/670/661330.pdf.
13. Loren Grush, "Elon Musk's Tesla Overshot Mars' Orbit, But It Won't Reach the Asteroid Belt as Claimed," *The Verge,* February 8, 2018, https://www.theverge.com/2018/2/6/16983744/spacex-tesla -falcon-heavy-roadster-orbit-asteroid-belt-elon-musk-mars.
14. Hamza Shaban, "Elon Musk Says He Will Probably Move to Mars," *Washington Post,* November 26, 2018, https://www.washingtonpost

.com/technology/2018/11/26/elon-musk-says-he-will-probably-move-mars.

15. Kenneth Chang, "Space Launch Firms Start Small Today to Go Big Tomorrow," *New York Times,* November 12, 2018, B2.
16. Gary Martin, "NewSpace: The 'Emerging' Commercial Space Industry," undated NASA presentation, circa 2014, https://ntrs.nasa.gov/archive/nasa/casi.ntrs.nasa.gov/20140011156.pdf.
17. NASA Fact Sheet, "Advanced Space Transportation Program: Paving the Highway to Space," 2008, https://www.nasa.gov/centers/marshall/news/background/facts/astp.html (retrieved March 21, 2018).
18. Sagdeev, *Making.*
19. W. David Compton and Charles D. Benson, *Living and Working in Space: A History of Skylab* (Washington, DC: NASA 1983), 271.
20. Compton and Benson, *Living and Working in Space,* 324.
21. Paul Martin, "Extending the Operational Life of the International Space Station Until 2024," NASA Office of Inspector General, Audit Report, IG-14-031, September 18, 2014.
22. Jeff Foust, "NASA Sees Strong International Interest in Lunar Exploration Plans," *Space News,* March 6, 2018, http://spacenews.com/nasa-sees-strong-international-interest-in-lunar-exploration-plans.
23. Robert M. Lightfoot Jr., "Return to the Moon: A Partnership of Government, Academia, and Industry," symposium, Washington, DC, March 28, 2018.
24. Eric Berger, "Former NASA Administrator Says Lunar Gateway Is 'a Stupid Architecture,'" *Ars Technica,* November 15, 2018, https://arstechnica.com/science/2018/11/former-nasa-administrator-says-lunar-gateway-is-a-stupid-architecture.
25. Mike Eckel, "First Female Space Tourist Blasts Off," Associated Press, September 18, 2006, http://news.yahoo.com/s/ap/20060918/ap_on_sc/russia_space.
26. Jeff Foust, "NASA Tries to Commercialize the ISS, Again," *Space Review,* June 10, 2019, http://www.thespacereview.com/article/3731/1.
27. Eric Niiler, "Who's Going to Buy the International Space Station?" *Wired,* February 12, 2018, https://www.wired.com/story/whos-going-to-buy-the-international-space-station.

28. Erin Mahoney, ed., "First Year of BEAM Demo Offers Valuable Data on Expandable Habitats," NASA, May 26, 2017, https://www.nasa.gov/feature/first-year-of-beam-demo-offers-valuable-data-on-expandable-habitats.
29. Dan Schrimpsher, "Interview: TransHab Developer William Schneider," *Space Review,* August 21, 2006, http://www.thespacereview.com/article/686/1.
30. Schrimpsher, "Interview."
31. Lara Logan (correspondent), interview with Robert Bigelow. *60 Minutes,* May 28, 2017, https://www.cbsnews.com/news/bigelow-aerospace-founder-says-commercial-world-will-lead-in-space/.
32. Robert Bigelow, "Public-Private Partnerships in Lunar Enterprises—No Time To Lose," presentation at the 2017 ISS R&D Conference, July 21, 2017, https://youtu.be/5403y2izgOo.
33. Paul Brians, "The Day They Tested the Rec Room," *CoEvolution Quarterly* (Summer 1981), 116–124.
34. "Reaction Engines Secures Funding to Enable Development of SABRE Demonstrator Engine," Reaction Engines Limited press release, July 12, 2016, https://www.reactionengines.co.uk/news/reaction-engines-secures-funding-to-enable-development-of-sabre-demonstrator-engine; "Reaction Engines Awarded DARPA Contract to Perform High-Temperature Testing of the SABRE Precooler," Reaction Engines Limited press release, September 25, 2017, https://www.reactionengines.co.uk/news/reaction-engines-awarded-darpa-contract-to-perform-high-temperature-testing-of-the-sabre-precooler.
35. Mike Wall, "Ticket Price for Private Spaceflights on Virgin Galactic's SpaceShipTwo Going Up," *Space,* April 30, 2013, https://www.space.com/20886-virgin-galactic-spaceshiptwo-ticket-prices.html.
36. Jeff Foust, "Still Waiting on Space Tourism after All These Years," *Space Review,* June 18, 2018, http://www.thespacereview.com/article/3516/1.
37. Federal Register, vol. 71, no. 241, December 15, 2006, p. 75616.
38. William M. Leary, "Robert Fulton's Skyhook and Operation Coldfeet," Center for the Study of Intelligence, Central Intelligence Agency,

June 27, 2008, https://www.cia.gov/library/center-for-the-study-of-intelligence/csi-publications/csi-studies/studies/95unclass/Leary.html.

39. Thomas Bogar et al., "Hypersonic Airplane Space Tether Orbital Launch System," NASA Institute for Advanced Concepts Research, Grant No. 07600-018, Phase I Final Report, http://images.spaceref.com/docs/spaceelevator/355Bogar.pdf.
40. John E. Grant, "Hypersonic Airplane Space Tether Orbital Launch—HASTOL," NASA Institute for Advanced Concepts 3rd Annual Meeting, NASA Ames Research Center, San Jose, California, June 6, 2001, http://www.niac.usra.edu/files/library/meetings/annual/jun01/391Grant.pdf.
41. Chia-Chi Chang et al., "A New Lower Limit for the Ultimate Breaking Strain of Carbon Nanotubes," *ACS Nano* 4 (2010): 5095–5100, doi:10.1021/nn100946q.
42. Paul Birch, "Orbital Ring Systems and Jacob's Ladders—II," *Journal of the British Interplanetary Society* 36 (1983): 231–238.
43. Doris Elin Salazar, "This Giant, Ultrathin NASA Balloon Just Broke an Altitude Record," *Space,* September 12, 2018, https://www.space.com/41791-giant-nasa-balloon-big-60-breaks-record.html.

4. Living on the Moon

1. Roger D. Launius, "Sputnik and the Origins of the Space Age," NASA https://history.nasa.gov/sputnik/sputorig.html (retrieved May 26, 2018).
2. Bob Allen, ed., "NASA Langley Research Center's Contributions to the Apollo Program," NASA Langley Research Center Fact Sheet, https://www.nasa.gov/centers/langley/news/factsheets/Apollo.html (retrieved November 11, 2018).
3. Launius, "Sputnik."
4. Zheng Wang, "National Humiliation, History Education, and the Politics of Historical Memory: Patriotic Education Campaign in China," *International Studies Quarterly* 52 (2008): 783–806.
5. Kevin Pollpeter et al., "China Dream, Space Dream: China's Progress in Space Technologies and Implications for the United States," report prepared for the U.S.-China Economic and Security Review Commission, 2015, https://www.uscc.gov/Research/china

-dream-space-dream-chinas-progress-space-technologies-and -implications-united-states.

6. Pollpeter et al., "China Dream, Space Dream."
7. GBTimes, "Lunar Palace-1: A Look Inside China's Self-Contained Moon Training Habitat," May 16, 2018, https://gbtimes.com/lunar -palace-1-a-look-inside-chinas-self-contained-moon-training-habitat.
8. Harrison Schmitt, "Return to the Moon: A Partnership of Government, Academia, and Industry," symposium, Washington, DC, March 28, 2018.
9. Christian Davenport, "Government Watchdog Says Cost of NASA Rocket Continues to Rise, a Threat to Trump's Moon Mission," *Washington Post,* June 18, 2019, https://www.washingtonpost.com /technology/2019/06/18/government-watchdog-says-cost-nasa -rocket-continues-rise-threat-trumps-moon-mission.
10. Davenport, "Government Watchdog."
11. Leonard David, "China's Anti-Satellite Test: Worrisome Debris Cloud Circles Earth," *Space,* February 2, 2007, https://www.space.com/3415 -china-anti-satellite-test-worrisome-debris-cloud-circles-earth.html.
12. Anne-Marie Brady, "China's Expanding Antarctic Interests: Implications for Australia," Australian Strategic Policy Institute, August 2017, https://www.aspi.org.au/report/chinas-expanding -interests-antarctica.
13. George F. Sowers, testimony, US House of Representatives Subcommittee on Space, Committee on Science, Space and Technology, September 7, 2017, https://www.hq.nasa.gov/legislative /hearings/9-7-17%20SOWERS.pdf.
14. Anthony Colaprete et al., "Detection of Water in the LCROSS Ejecta Plume," *Science* 330 (2010): 463–468, doi:10.1126/ science.1186986; Shuai Li et al., "Direct Evidence of Surface Exposed Water Ice in the Lunar Polar Regions," *PNAS* 115 (2018), doi:10.1073/pnas.1802345115.
15. Mike Wall, "Mining the Moon's Water: Q&A with Shackleton Energy's Bill Stone," *Space,* January 13, 2011, https://www.space .com/10619-mining-moon-water-bill-stone-110114.html.
16. Harrison Schmitt, *Return to the Moon: Exploration, Enterprise, and Energy in the Human Settlement of Space* (Göttingen, Germany: Copernicus, 2006).

17. Schmitt, *Return to the Moon.*
18. Ian A. Crawford, "Lunar Resources: A Review," *Progress in Physical Geography: Earth and Environment* 39 (2015): 137–167, doi:10.1177/0309133314567585.
19. Amanda Kay, "Rare Earths Production: 8 Top Countries," *Investing News,* April 3, 2018, https://investingnews.com/daily/resource-investing/critical-metals-investing/rare-earth-investing/rare-earth-producing-countries.
20. Angel Abbud-Madrid, private communication, Return to the Moon symposium, March 28, 2018.
21. Alex Freundlich et al., "Manufacture of Solar Cells on the Moon," Conference Record of the Thirty-First IEEE Photovoltaic Specialists Conference, 2005, doi:10.1109/PVSC.2005.1488252.
22. David Biello, "Where Did the Carter White House's Solar Panels Go?" *Scientific American,* August 6, 2010, https://www.scientificamerican.com/article/carter-white-house-solar-panel-array.
23. Haym Benaroya et al., "Engineering, Design and Construction of Lunar Bases," *Journal of Aerospace Engineering* 15 (2002): 33–45.
24. Benaroya et al., "Engineering."
25. Kenneth Chang, "NASA Reports a Moon Oasis, Just a Little Bit Wetter Than the Sahara" *New York Times,* October 22, 2010, A20.
26. Li et al., "Direct Evidence."
27. Crawford, "Lunar Resources," 167.
28. Cheryl Lynn York et al., "Lunar Lava Tube Sensing," Lunar and Planetary Institute, Joint Workshop on New Technologies for Lunar Resource Assessment, 1992, https://www.lpi.usra.edu/lpi/contribution_docs/TR/TR_9206.pdf.
29. Tetsuya Kaku et al., "Detection of Intact Lava Tubes at Marius Hills on the Moon by SELENE (Kaguya) Lunar Radar Sounder," *Geophysical Research Letters* 44 (2017): 10,155–10,161, doi:10.1002/2017GL074998.
30. Werner Grandl, "Human Life in the Solar System," *REACH* 5 (2017): 9–21, doi:10.1016/j.reach.2017.03.001.
31. Junichi Haruyama et al., "Lunar Holes and Lava Tubes as Resources for Lunar Science and Exploration," in *Moon—*

Prospective Energy and Material Resources, ed. Viorel Badescu (Springer, 2012), 139–163.

32. Gerald B. Sanders and William E. Larson, "Progress Made in Lunar In Situ Resource Utilization under NASA's Exploration Technology and Development Program," *Journal of Aerospace Engineering* 26 (2013), doi:10.1061/(ASCE)AS.1943-5525.0000208.
33. "NASA's Analog Missions: Paving the Way for Space Exploration," NASA fact sheet, NP-2011-06-395-LaRC, 2011, https://www.lpi.usra.edu/lunar/strategies/NASA-Analog-Missions-NP-2011-06-395.pdf.
34. Kenneth Chang, "Meet SpaceX's First Moon Voyage Customer, Yusaku Maezawa," *New York Times,* September 12, 2018, B4.
35. Rachel Caston et al., "Assessing Toxicity and Nuclear and Mitochondrial DNA Damage Caused by Exposure of Mammalian Cells to Lunar Regolith Simulants," *GeoHealth* 2 (2018): 139–148, doi:10.1002/2017GH000125.
36. "Apollo 17 Technical Crew Debriefing," NASA, January 4, 1973, http://www.ccas.us/CCAS_NASA_PressKits/Apollo_Missions/Apollo17_TechnicalCrewDebriefing.pdf.
37. John T. James and Noreen Kahn-Mayberry, "Risk of Adverse Health Effects from Lunar Dust Exposure," in *Human Health and Performance Risks of Space Exploration Missions,* ed. Jancy C. McPhee and John B. Charles (Washington, DC: NASA, 2009), 317–330.
38. G. W. Wieger Wamelink et al., "Can Plants Grow on Mars and the Moon: A Growth Experiment on Mars and Moon Soil Simulants," *PLoS ONE* 9 (2014), doi:10.1371/journal.pone.0103138.
39. Matt Williams, "How Do We Terraform The Moon?" *Universe Today,* March 31, 2016, https://www.universetoday.com/121140/could-we-terraform-the-moon.
40. Tariq Malik, "The Moon Will Get Its Own Mobile Phone Network in 2019," *Space,* February 28, 2018 https://www.space.com/39835-moon-mobile-phone-network-ptscientists-2019.html.
41. Astronomer-author Phil Plait provides this logic and other elements of debunking in his 2002 book *Bad Astronomy* (New York: John Wiley).

5. Living on Asteroids

1. Stephen P. Maran, *Astronomy for Dummies,* 4th ed. (Hoboken, NJ: John Wiley & Sons, 2017).
2. Brid-Aine Parnell, "NASA Will Reach Unique Metal Asteroid Worth $10,000 Quadrillion Four Years Early," *Forbes,* May 26, 2017, https://www.forbes.com/sites/bridaineparnell/2017/05/26/nasa-psyche-mission-fast-tracked.
3. Giovanni Bignami and Andrea Sommariva, *The Future of Human Space Exploration* (London: Palgrave Macmillan, 2016).
4. Kenneth Chang, "The Osiris-Rex Spacecraft Begins Chasing an Asteroid," *New York Times,* September 9, 2016, A12.
5. "NASA's OSIRIS-REx Asteroid Sample Return Mission," NASAfacts, FS-206-4-411-GSFC, NASA, May 2016, https://www.nasa.gov/sites/default/files/atoms/files/osiris_rex_factsheet5-25.pdf.
6. Detlef Koschny, ESA, private communication, June 30, 2018.
7. Axel Hagermann, Hayabusa2 team member, private communication, June 29, 2018.
8. Jeff Foust, "Asteroid Mining Company Planetary Resources Acquired by Blockchain Firm," *SpaceNews,* October 31, 2018, https://spacenews.com/asteroid-mining-company-planetary-resources-acquired-by-blockchain-firm.
9. Martin Elvis, "How Many Ore-Bearing Asteroids?" *Planetary and Space Science* 91 (2014): 20–26, doi:10.1016/j.pss.2013.11.008.
10. Martin Elvis, private communication, June 28, 2018.
11. Kenneth Chang, "If No One Owns the Moon, Can Anyone MakeMoney Up There?" *New York Times,* November 18, 2017, D1.
12. Jeff Foust, "Luxembourg Adopts Space Resources Law," *Space News,* July 17, 2017, http://spacenews.com/luxembourg-adopts-space-resources-law.
13. Andrew Zaleski, "Luxembourg Leads the Trillion-Dollar Race to Become the Silicon Valley of Asteroid Mining," CNBC, April 16, 2018, https://www.cnbc.com/2018/04/16/luxembourg-vies-to-become-the-silicon-valley-of-asteroid-mining.html.
14. Thomas Prettyman et al., "Extensive Water Ice within Ceres' Aqueously Altered Regolith: Evidence from Nuclear

Spectroscopy," *Science* 355 (2017): 55–59, doi:10.1126/science.aah6765; Maria Cristina De Sanctis et al., "Localized Aliphatic Organic Material on the Surface of Ceres," *Science* 355 (2017): 719–722, doi:10.1126/science.aaj2305.

15. Werner Grandl, "Human Life in the Solar System," *REACH—Reviews in Human Space Exploration* 5 (2017): 9–21, doi:10.1016/j.reach.2017.03.001.
16. Werner Grandl, private communication, July 4, 2018.
17. Peter A. Curreri and Michael K. Detweiler, "A Contemporary Analysis of the O'Neill—Glaser Model for Space-based Solar Power and Habitat Construction," *NSS Space Settlement Journal,* December 2011.

6. Living on Mars

1. "NASA's Journey to Mars: Pioneering Next Steps in Space Exploration," NASA NP-2015-08-2018-HQ, October 2015, https://www.nasa.gov/sites/default/files/atoms/files/journey-to-mars-next-steps-20151008_508.pdf.
2. George W. Bush, "A Renewed Spirit of Discovery: The President's Vision for U.S. Space Exploration," White House fact sheet, January 2004, https://permanent.access.gpo.gov/lps72574/renewed_spirit.pdf.
3. John M. Logsdon, "Ten Presidents and NASA," in *50th Magazine—50 Years of Exploration and Discovery,* NASA (2008), https://www.nasa.gov/50th/50th_magazine/10presidents.html.
4. Kenneth Chang, "NASA Budgets for a Trip to the Moon, but Not While Trump Is President," *New York Times,* February 12, 2018, A13; Donald Trump, "Presidential Memorandum on Reinvigorating America's Human Space Exploration Program," US Presidential Memorandum, December 11, 2017, https://www.whitehouse.gov/presidential-actions/presidential-memorandum-reinvigorating-americas-human-space-exploration-program; Jeff Foust, "Space Force? Create a 'Space Guard' Instead, Some Argue," *Space News,* May 31, 2018, http://spacenews.com/space-force-create-a-space-guard-instead-some-argue.
5. Associated Press, "Moon Landing a Big Waste, Says Barry," *Tuscaloosa News,* September 21, 1963, 8.

6. Charles D. Hunt and Michel O. Vanpelt, "Comparing NASA and ESA Cost Estimating Methods for Human Missions to Mars," 26th International Society of Parametric Analysts Conference, Frascati, Italy, May 10–12, 2004, https://ntrs.nasa.gov/archive/nasa/casi.ntrs.nasa.gov/20040075697.pdf.
7. Irene Klotz, "NASA Looking to Mine Water on the Moon and Mars," *Space News,* January 28, 2014, https://spacenews.com/39307nasa-planning-for-mission-to-mine-water-on-the-moon.
8. Gerald B. Sanders and William E. Larson, "Progress Made in Lunar In Situ Resource Utilization under NASA's Exploration Technology and Development Program," *Journal of Aerospace Engineering* 26 (2013), doi:10.1061/(ASCE)AS.1943-5525.0000208.
9. Laurie Chen, "China's Mars Base Plan Revealed . . . and Covering 95,000 sq km, There's Certainly Plenty of Space," *South China Morning Post,* September 7, 2017, https://www.scmp.com/news/china/society/article/2110051/there-price-mars-chinas-red-planet-simulator-set-cost-us61.
10. Thor Hogan, "Lessons Learned from the Space Exploration Initiative," NASA History Division, *News & Notes* 24, no. 4 (November 2007).
11. Thor Hogan, *Mars Wars: The Rise and Fall of the Space Exploration Initiative* (Washington, DC: NASA History Series SP-2007-4410, 2007), 2.
12. Robert Zubrin, email correspondence, August 20, 2018.
13. Dai Shiba et al., "Development of New Experimental Platform 'MARS'—Multiple Artificial-Gravity Research System—To Elucidate the Impacts of Micro / Partial Gravity on Mice," *Scientific Reports* 7 (2017), doi:10.1038/s41598-017-10998-4.
14. Michael J. Carlowicz and Ramon E. Lopez, *Storms from the Sun* (Washington, DC: Joseph Henry Press, 2002), 144.
15. Carlowicz and Lopez, *Storms,* 144.
16. T. Troy McConaghy et al., "Analysis of a Class of Earth-Mars Cycler Trajectories," *Journal of Spacecraft and Rockets* 41 (2004): 622–628, doi:10.2514/1.11939.
17. George Schmidt et al., "Nuclear Pulse Propulsion: Orion and Beyond," 36th AIAA / ASME / SAE / ASEE Joint Propulsion

Conference & Exhibit, July 16–19, 2000, Huntsville, Alabama, https://ntrs.nasa.gov/search.jsp?R=20000096503.

18. Jeff Foust, "Review: Mars One: Humanity's Next Great Adventure," *Space Review,* March 14, 2016, htp://www.thespacereview.com /article/2940/1.
19. Bret G. Drake and Kevin D. Watts, eds., "Human Exploration of Mars Design Reference Architecture 5.0, Addendum #2," NASA / SP–2009–566-ADD2, NASA Headquarters, March 2014, https://www .nasa.gov/sites/default/files/files/NASA-SP-2009-566-ADD2.pdf.
20. Donald M. Hassler et al., "Mars' Surface Radiation Environment Measured with the Mars Science Laboratory's Curiosity Rover," *Science* 343 (2014), doi:10.1126/science.1244797.
21. Robert W. Moses and Dennis M. Bushnell, "Frontier In-Situ Resource Utilization for Enabling Sustained Human Presence on Mars," NASA / TM–2016-219182, April 2016, https://ntrs.nasa.gov /search.jsp?R=20160005963.
22. Tobias Owen et al., "The Composition of the Atmosphere at the Surface of Mars," *Journal of Geophysical Research* 82 (1977): 4635–4639, doi:10.1029/JS082i028p04635.
23. Christopher McKay, private communication, July 22–23, 2018.
24. Christopher McKay et al., "Utilizing Martian Resources for Life Support," in *Resources of Near-Earth Space,* ed. John S. Lewis, Mildred Shapley Matthews, and Mary L. Guerrieri (Tucson: University of Arizona Press, 1993), 819–843.
25. Alfonso F. Davila at al., "Perchlorate on Mars: A Chemical Hazard and a Resource for Humans," *International Journal of Astrobiology* 12 (2013): 321–325.
26. Edward Guinan, private communication, 231st Meeting of the American Astronomical Society, Washington, DC, January 8, 2018.
27. Christopher Wanjek, "Ground Control to 'The Martian': Good Luck with Them Potatoes," *Live Science,* October 9, 2015, https://www .livescience.com/52438-the-martian-potatoes-health-effects.html.
28. Alexandra Witze, "There's Water on Mars! Signs of Buried Lake Tantalize Scientists," *Nature* 560 (2018): 13–14, doi:10.1038/ d41586-018-05795-6.

29. Robert Zubrin, *The Case for Mars: The Plan to Settle the Red Planet and Why We Must* (New York: Free Press, 2011), 187–232.
30. Zubrin, *Case for Mars,* 246–247.
31. Adam E. Jakus et al., "Robust and Elastic Lunar and Martian Structures from 3D-Printed Regolith Inks," *Scientific Reports* 7 (2017), doi:10.1038/srep44931.
32. William Harwood, "Curiosity Relies on Untried 'Sky Crane' for Descent to Mars," *CBS News,* July 30, 2012, http://www.cbsnews.com/network/news/space/home/spacenews/files/msl_preview_landing.html.
33. Kasandra Brabaw, "MIT Team Wins Mars City Design Contest for 'Redwood Forest' Idea," *Space,* November 25, 2017, https://www.space.com/38881-mit-team-wins-mars-city-design-competition.html.
34. William J. Rowe, "The Case for an All-Female Crew to Mars," *Journal of Men's Health & Gender* 1 (2004): 341–344, doi:10.1016/j.jmhg.2004.09.006.
35. Lauren Blackwell Landon at al., "Selecting Astronauts for Long-Duration Exploration Missions: A Retrospective Review and Considerations for Team Performance and Functioning," NASA TM-2016-219283, December 1, 2016.
36. Freeman Dyson, *Disturbing the Universe* (New York: Harper & Row, 1979), 118–126.
37. NSF OPP Budget Request to Congress FY2019, https://www.nsf.gov/about/budget/fy2019/pdf/30_fy2019.pdf.
38. Scott Solomon, "The Martians Are Coming—and They're Human" *Nautilus,* October 27, 2016, http://nautil.us/issue/41/selection/the-martians-are-comingand-theyre-human.
39. Bruce M. Jakosky and Christopher S. Edwards, "Inventory of CO2 Available for Terraforming Mars," *Nature Astronomy* 2 (2018):634–639, doi:10.1038/s41550-018-0529-6.
40. Partha P. Bera et al., "Design Strategies to Minimize the Radiative Efficiency of Global Warming Molecules," *PNAS* 107 (2010): 9049–9054, doi:10.1073/pnas.0913590107.
41. Zubrin, *Case for Mars,* 269–270.
42. Margarita M. Marinova et al., "Radiative-Convective Model of Warming Mars with Artificial Greenhouse Gases," *Journal of Geophysical Research* 110 (2005), doi:10.1029/2004JE002306.

7. Living in the Inner and Outer Solar System and Beyond

1. Stephen Maran, *Astronomy for Dummies,* 4th ed., (Hoboken, NJ: John Wiley & Sons, 2017), 121–122.
2. Geoffrey A. Landis et al., "Atmospheric Flight on Venus," 40th Aerospace Sciences Meeting and Exhibit, American Institute of Aeronautics and Astronautics, Reno, Nevada, January 14–17, 2002, NASA / TM—2002-211467, https://www.researchgate.net /publication/24286050_Atmospheric_Flight_on_Venus.
3. Paul Birch, "Terraforming Venus Quickly," *Journal of the British Interplanetary Society* 44 (1991): 157–167.
4. Mark Bullock and David H. Grinspoon, "The Stability of Climate on Venus," *Journal of Geophysical Research* 101 (1996): 7521–7529, doi:10.1029/95JE03862.
5. Robert Zubrin, *The Case for Space: How the Revolution in Spaceflight Opens Up a Future of Limitless Possibility* (New York: Prometheus Books, 2019), 166.
6. Zubrin, *Case for Space,* 166.
7. John M. Wahr et al., "Tides on Europa, and the Thickness of Europa's Icy Shell," *Journal of Geophysical Research* 111 (2006): 12005–12014, doi:10.1029/2006JE002729.
8. Zubrin, *Case for Space,* 166.
9. Zubrin, *Case for Space,* 166.
10. Christopher McKay et al., "The Possible Origin and Persistence of Life on Enceladus and Detection of Biomarkers in the Plume," *Astrobiology* 8 (2008): 909–919, doi:10.1089/ast.2008.0265.
11. J. Hunter Waite et al., "Cassini Finds Molecular Hydrogen in the Enceladus Plume: Evidence for Hydrothermal Processes," *Science* 356 (2017): 155–159, doi:10.1126/science.aai8703.
12. Charles Wohlforth and Amanda R. Hendrix, *Beyond Earth: Our Path to a New Home in the Planets* (New York: Pantheon, 2016).
13. James Stevenson et al., "Membrane Alternatives in Worlds without Oxygen: Creation of an Azotosome," *Science Advances* 1 (2015), 1:e1400067, doi:10.1126/sciadv.1400067.
14. Isaac Arthur, "Outward Bound: Colonizing Titan," *Science & Futurism with Isaac Arthur,* October 12, 2017, https://www.youtube .com/watch?v=HdpRxGjtCo0&vl=en.

15. Terry A. Hurford et al., "Triton's Fractures as Evidence for a Subsurface Ocean," 48th Lunar and Planetary Science Conference, March 20–24, 2017, The Woodlands, Texas, https://www.hou.usra.edu/meetings/lpsc2017/pdf/2376.pdf.
16. S. Alan Stern, "The Pluto System: Initial Results from Its Exploration by New Horizons," *Science* 350 (2015): 292, doi:10.1126/science.aad1815.
17. Leonid Marochnik et al., "Estimates of Mass and Angular Momentum in the Oort Cloud," *Science* 242 (1988): 547–550, doi:10.1126/science.242.4878.547.
18. "How Do We Know When Voyager Reaches Interstellar Space?" NASA fact sheet 2013-278, September 12, 2013, https://www.nasa.gov/mission_pages/voyager/voyager20130912f.html.
19. Isaac Arthur, "Outward Bound: Colonizing the Oort Cloud," *Science and Futurism with Isaac Arthur,* December 14, 2017, https://www.youtube.com/watch?v=H8Bx7y0syxc.
20. Andrew V. Ilin et al., "VASIMR® Human Mission to Mars," Space, Propulsion & Energy Sciences International Forum, March 15–17, 2011, University of Maryland, College Park, MD, http://www.adastrarocket.com/Andrew-SPESIF-2011.pdf.
21. Martin Tajmar et al., "The SpaceDrive Project—First Results on EMDrive and Mach-Effect Thrusters," presented at the Space Propulsion Conference in Seville, Spain, May 14–18, 2018, https://www.researchgate.net/publication/325177082_The_SpaceDrive_Project_-_First_Results_on_EMDrive_and_Mach-Effect_Thrusters.
22. Gregory L. Matloff et al., "The Beryllium Hollow-Body Solar Sail: Exploration of the Sun's Gravitational Focus and the Inner Oort Cloud," Cornell University Physics, September 20, 2008, arXiv:0809.3535.

Additional Reading and Listening

Videos and podcasts

Answers with Joe, hosted by Joe Scott, https://www.youtube.com/channel/UC-2YHgc363EdcusLIBbgxzg

Astronomy Cast, hosted by Fraser Cain and Pamela Gay, http://www.astronomycast.com/

Science & Futurism with Isaac Arthur, hosted by Isaac Arthur, https://www.youtube.com/channel/UCZFipeZtQM5CKUjx6grh54g

StarTalk, hosted by astrophysicist Neil deGrasse Tyson, https://www.startalkradio.net/

Veritasium, hosted by Derek Muller, https://www.youtube.com/channel/UCHnyfMqiRRG1u-2MsSQLbXA

Online publications

Bad Astronomy, blog by astronomer Phil Plait, http://www.badastronomy.com/index.html

New York Times space calendar app, https://www.nytimes.com/interactive/2017/science/astronomy-space-calendar.html

Space.com, a space and astronomy news website

Space Review, in-depth articles and commentary regarding all aspects of space exploration, edited by Jeff Foust, http://www.thespacereview.com/

Universe Today, published by Fraser Cain, https://www.universetoday.com/

Fiction

The Fountains of Paradise, science fiction by Arthur C. Clarke (New York: Harcourt Brace Jovanovich, 1979)

The Martian Way, science fiction novella by Isaac Asimov (Garden City, NY: Doubleday, 1955)

Red Mars, Green Mars, and *Blue Mars,* a trilogy by Kim Stanley Robinson (New York: Bantam Books, 1993, 1994, 1996)

Rendezvous with Rama, science fiction by Arthur C. Clarke (New York: Harcourt Brace Jovanovich, 1973)

Nonfiction

The Case for Mars: The Plan to Settle the Red Planet and Why We Must, by Robert Zubrin (New York: Free Press, 2011)

The Case for Space: How the Revolution in Spaceflight Opens Up a Future of Limitless Possibility, by Robert Zubrin (Amherst, NY: Prometheus, 2019)

Disturbing the Universe, a collection of essays by Freeman Dyson (New York: Harper and Row, 1979)

The Future of Humanity: Terraforming Mars, Interstellar Travel, Immortality, and Our Destiny beyond Earth, by Michio Kaku (New York: Doubleday, 2018)

The High Frontier: Human Colonies in Space, by Gerard K. O'Neill (New York: Morrow, 1976)

Packing for Mars: The Curious Science of Life in the Void, by Mary Roach (New York: Norton, 2010)

Pale Blue Dot: A Vision of the Human Future in Space, by Carl Sagan (New York: Random House, 1994)

Space Chronicles: Facing the Ultimate Frontier, a collection of commentary by Neil deGrasse Tyson (New York: Norton, 2012)

Vacation Guide to the Solar System: Science for the Savvy Space Traveler!, a "fictional" nonfiction science book by Olivia Koski and Jana Grcevich of the Intergalactic Travel Bureau (New York: Penguin Random House, 2017)

Space societies

Mars Society, founded by Robert Zubrin, Chris McKay, and many others, with a focus on Mars settlements, https://www.marssociety.org/

National Space Society, a merger of the L5 Society, founded by Gerard O'Neill, and National Space Institute, founded by Wernher von Braun, with a focus on living and working in space, https://space.nss.org/

Planetary Society, founded by Carl Sagan and others, the largest space society and an advocate of robotic exploration of the solar system, http://www.planetary.org/

Space Frontier Foundation, an organization promoting private enterprise in space, http://newspace.spacefrontier.org/

Acknowledgments

A special thanks must go to Stephen Maran, my mentor at NASA Goddard Space Flight Center, who believed in my talent, inspired me with his wit and insight, and allowed me to grow as a writer.

I would also like to thank the following people who took the time via phone, email, or face-to-face interviews to explain to me their research or who otherwise aided in my writing of this book: Angel Abbud-Madrid, Irwin Arias, Isaac Arthur, Charles L. Baker, Tristan Bassingthwaighte, Nadia Biassou, Bill Branson, Ernie Branson, David Brin, Fraser Cain, Patrick Carroll, Laura Carter, Peter Checchia, Peter Curreri, Jeff Dean, Martin Elvis, Gene Giacomelli, Werner Grandl, Edward Guinan, Chris Gunn, Axel Hagermann, Keith Jarrett, Detlef Koschny, Christopher McKay, Marc Millis, Phil Plait, Barry Pryor, Alan Robock, Phil Sadler, Roald Sagdeev, Harrison Schmitt, Joe Scott, KC Shasteen, Isaac Silvera, Brian Toon, Carl Wanjek, Bingxin Yu, and E. Paul Zehr.

Illustration Credits

Page

24 ESO / M. Kornmesser / CC BY 4.0

28 shrimpo1967 / Wikimedia Commons / CC BY-SA 2.0

37 Dave Pape / Wikimedia Commons

46 NASA

61 DrStarbuck / Wikimedia Commons / CC BY 2.0

89 Zhihua Yang et al., "Radiation-Induced Brain Injury after Radiotherapy for Brain Tumor," *IntechOpen,* March 25, 2015, doi:10.5772/59045, figure 1C / CC BY 3.0 / © the Authors

113 SpaceX / Wikimedia Commons / CC0 1.0

118 NASA / Don Davis

126 NASA / Bill Ingalls

135 Virgin Galactic / Mark Greenberg / Wikimedia Commons // CC BY-SA 3.0

137 SpaceportAmerica

143 NASA / Pat Rawling

178 © ESA / Foster + Partners

188 NASA / Bill Anders

196 The Prototype Lunar Greenhouse (LGH) / Controlled Environment Agriculture Center, University of Arizona

207 JAXA

221 NASA / JPL-Caltech

254 NASA / Clouds AO / SEArch

293 NASA / JPL-Caltech / MSSS / TAMU

306 NASA / JPL-Caltech

311 NASA / SACD

324 NASA / JPL-Caltech / ASI / Cornell

Index

Abbud-Madrid, Angel, 170
Ad Astra Rocket Company, 339
Agreement Governing the Activities of States on the Moon and Other Celestial Bodies (Moon Treaty), 159, 215
Aldrin, Buzz, 186–187, 242, 246–247
Aldrin Mars Cycler, 242
Allen, John, 57
All-Terrain Hex-Limbed Extra-Terrestrial Explorer, 227
Almaz space station, 119
Alpha Centauri, 103, 244, 337, 341
aluminum, 163, 165, 169–170
aluminum-ice (ALICE), 104–105
ammonia, 203, 205, 211, 228, 328–329
Amundsen–Scott South Pole Station, 34–35
Ansari, Anousheh, 124
Ansari X Prize, 134
Anserlian, Garo, 286
Antarctica, 27, 29–37, 44
Antarctic Treaty, 31, 36, 160
antimatter, 116, 341–342
Apollo 1, 3, 8
Apollo 10, 243
Apollo 11, 4, 190, 223
Apollo 13, 4, 6, 8
Apollo 14, 84
Apollo 16, 6, 83
Apollo 17, 6, 83
Apollo program, 152–153
appendicitis, 92–93
aquaponics, 261
"areology," 287
argon, 258–259
Ariane 5 rocket, 113
Armstrong, Neil, 4, 186
Armstrong limit, 299, 301
Army Ballistic Missile Agency, 107
Arthur, Isaac, 326–327
artificial gravity. *See* centrifugal force
artificial intelligence (AI), 272
Asimov, Isaac, 203, 282, 326

Asteroid Redirect Mission (ARM), 211–212
asteroids: as habitats, 21, 217–218; defined, 201–204; impacts on Earth, 22–23; mining on, 5, 209–212; near-earth, 201, 206–207; potential profit from, 209–215; as starships, 218; water on, 205
Atlas rockets, 109, 113
atmospheric pressure, 41, 251, 297
augmented reality, 272–273
autonomous vehicles, 271
azotosome, 326

Baker, David, 223, 234
Bannon, Steve, 59
Bass, Ed, 57, 60
Bassingthwaighte, Tristan, 278
Bean, Alan, 238–239
Bennu asteroid, 206
Berezovoy, Anatoly, 45
Bezos, Jeff, 136
BFR, SpaceX rocket, 109–110
Bigelow, Robert, 125, 126–127
Bigelow Aerospace, 77, 114, 125–126, 129
Bigelow Expandable Activity Module (BEAM), 125–126
bioregenerative life support system (BLSS), 195–198, 260
Biosphere 2, 56–61, 275
Blue Origin, 114, 123, 136
bone loss, 68–69
boron nitride nanotubes (BNNTs), 91
brain damage. *See* cognitive health risks
Branson, Richard, 135
Braun, Wernher von, 99, 107, 150
Breakthrough Propulsion Physics Project, 341
Brin, David, 108n
Buran space shuttle, 134
Bush, George H. W., 222, 230–231
Bush, George W., 223

Callisto, 314, 318–319
Canadian Space Agency, 120, 228
carbon nanotubes, 142–143
Carter, Jimmy, 170, 222
Case for Mars, The (book), 227, 270
Case for Mars, The (conference), 259n
Cassini spacecraft, 320–321
centrifugal force, 73–74, 236–238
Centrifuge Accommodations Module (CAM), 76–77
Ceres, 201–203, 208, 215–218
CERN, 92
Cernan, Eugene, 148, 174, 176, 194
Chaffee, Roger, 3
Chandrayaan-1 mission, 162, 180
Chang Díaz, Franklin, 339
Charon, 331–332
Chelomey, Vladimir, 119
China: asteroid mining by, 213; Mars plans of, 220, 230, 249, 276; National Space Administration (CNSA), 154; rare-earth elements from, 168; space race by, 13, 123, 154–156; space stations of, 121–122; US exclusion policy against, 158
circulatory system, 69–70
climate change, 23–25
Clinton, Bill, 222
cobalt, 204, 211

cochlear implants, 272–273
cognitive health risks, 70, 85–89
colonization, definition of, 32
comets, 63, 202, 333–334; as starships, 334
Comité Speciale de l'Année Geophysique Internationale (CSAGI), 149
Commercial Space Launch Amendments Act of 2004, 136
Concorde, 138–139
Concordia Research Station, 48
ConsenSys, 208–209
Controlled Environment Agriculture Center (CEAC), 35–36, 195
Cook, Frederick A., 54–55
coronal mass ejection (CME), 80–81
cosmic rays. *See* radiation: cosmic
cosmic ray visual phenomena, 86
Cox Report, 158
cryptocurrency, 209
CubeSat, 114n
Cucinotta, Francis, 83, 88–89
Curiosity rover, 239, 251, 259, 272, 274

DARPA (US Defense Advanced Research Projects Agency), 133
Dawn spacecraft, 208, 216, 339
Dear Moon project, 186
Deep Space Gateway. *See* Lunar Orbital Platform-Gateway
Deep Space Industries (DSI), 209, 213
Deimos, 291
Delta rockets, 109, 113–114
delta-v (Δv), 99–101, 209
Desert RATS, 227–228
deuterium, 166–167, 268
Devon Island, Canada, 226–227, 261
dolphin (US Navy pin), 43–44
Dragonfly mission, 307, 325
Dragon spacecraft, 109, 125
Dyson, Freeman, 103–104, 242–244, 280, 333
dysprosium, 162

Earth–Mars cycler, 241–242, 246
Earthrise, 188
EDEN ISS, 36
Einstein, Albert, 99
Eisenhower, Dwight D., 107, 149–150
electrolysis, 40–41, 180
electrostatic levitation, 194
Elvis, Martin, 209
EmDrive, 340
Enceladus, 33, 314, 320–321
escape velocity, 164, 212–213
Esperanza Base, 32
Europa, 33, 39, 314, 316–317
European Space Agency (ESA): International Space Station, 120; Mars activities, 52, 276; Moon activities, 176–179; rocketry, 113; Space Medicine Office, 69; Titan missions, 321, 324
European Space Radiation Superconducting Shield (SR2S) project, 92; exercise in space, 69
exhaust velocity, 100, 104
ExoMars Orbiter / Schiaparelli EDL Demo Lander mission, 276
Explorer 1 satellite, 150
Extravehicular Mobility Units (EMU), 251–252
eye health, 70

Falcon Heavy rocket, 112–113, 115
Falcon rocket family, 109–113
Federal Aviation Administration (FAA), 136, 138
Fine Motor Skills investigation, 71
fission, nuclear, 105–106, 342
Flashline Mars Arctic Research Station (FMARS), 226–227
Fluid Shifts investigation, 71
free-fall, 67, 99
free-return trajectory, 242
Fulton surface-to-air recovery system (STARS), 139
Functional Task investigation, 71
funding, motivations for, 11–15
fusion, nuclear, 105–106, 165–168, 334

gadolinium, 162
gamma-ray bursts, 25
gamma rays, 18
Ganymede, 317–318, 320
generational starship, 337
geostationary orbit, 127, 142–143
geosynchronous orbit, 127
g-force, 243–244
Giacomelli, Gene, 195, 197
Gibson, Edward, 119
gold, 202, 204–205, 268–269
Goldin, Dan, 232
Goldwater, Barry, 223
gravity: defined, 67–68; effects on health, 66–78, 173
Griffin, Michael, 108–109, 123
Grissom, Gus, 3
Groton, CT, submarine base, 44, 46
Guinan, Edward, 262

Hall thruster, 339
Haughton Mars Project (HMP), 227–228
Hawai'i Space Exploration Analog and Simulation (HI-SEAS), 48–51, 226, 278
Hawking, Stephen, 25–26
Hayabusa probes, 207–208
health risks of space travel, 64–66
heliocentric orbit, 151
helium-3, 161, 165–168, 181, 268
Hermes (fictional spacecraft), 236
Hess, Victor Franz, 86
Hohmann transfer orbit, 241
Hubble Space Telescope, 127, 231
Human Exploration Research Analog (HERA), 45–47, 226
Human Health and Performance Directorate, NASA, 96
Human Outer Planets Exploration (HOPE), 318
Huygens mission, 321
hydrazine, 104
hydroponics, 36, 195–198, 260–261
hypercapnia, 95
Hypersonic Airplane Space Tether Orbital Launch (HASTOL) system, 140–142
hypoxia, 95

Iapetus, 319–320
IceCube Neutrino Observatory, 33
Ice Home, 253–255
IKAROS spacecraft, 340
immunodeficiency, 93, 96
inflatable habitats, 77, 125–126, 129, 176–177, 253, 255
InSight probe, 239

in situ resource utilization (ISRU), 185, 228–230
intergalactic highway, 336
International Geophysical Year, 149
International Space Station (ISS): cost of, 10, 290; microgravity research on, 66–69, 75; modules attached to, 76–77; orbit of, 26; origin of, 120–121; tourism and, 124–126, 128
interstellar ark, 337
Io, 315–316
ionizing radiation, 79–81
ion propulsion, 207–208, 338–339
iron, 163, 169–170, 331
Italian Space Agency, 265
Itokawa asteroid, 207

Japan Aerospace Exploration Agency (JAXA), 207–208, 338, 340
Japan National Space Development Agency, 76
Johnson, Lyndon B., 5, 153, 223
Jupiter, 305–306, 314–319
Jupiter-3 rocket, 150

Kaluga, Russia, 98–99
Kármán line, 134–135, 146
Kelly, Mark, 94
Kelly, Scott, 94–95, 235, 238
Kennedy, John F., 5–6, 153
Kerimov, Kerim, 119
Kevlar, 126, 140, 145, 269, 274
Khrushchev, Nikita, 149
Krauss, Lawrence, 246–247
KREEP, 162, 181
Kuiper belt, 305, 330–333
Kursk (Russian submarine), 42

Lagrangian points, 127
Laika, the dog, 150
Lake Vostok, 33, 316
Landscape Evolution Observatory, 60
lava tubes, 184–185
Lebedev, Valentin Vitalyevich, 45
Lee, Pascal, 228
Lehrer, Tom, 150n
Lewis and Clark expedition, 205–206
L5 Society, 159
Lightfoot, Robert, Jr., 122
Limoli, Charles, 87
liquid oxygen (LOX), 104
Logsdon, John, 5–6
Long March rockets, 113–114, 155
lonsdaleite, 205
Lopez-Alegria, Michael, 124
Low-Density Supersonic Decelerator (LDSD), 274
low-earth orbit (LEO), 127
LUCA ("loud, unfriendly counter-arguments"), 112, 171, 277, 330
Lunar Hilton, 186–187
lunar hopper, 192
Lunar Legacy Project, 190
lunar-Mars greenhouse. *See* bioregenerative life support system
Lunar Orbital Platform-Gateway (LOP-G), 122–123
Lunar Palace 1, 155, 197
Lunar Reconnaissance Orbiter, 185
Lunar Roving Vehicles, 191
Luna spacecraft, 151–152
Luxembourg, 213–214

Maezawa, Yusaku, 186
magnesium, 163, 165
magnetosphere, 18, 80, 128

Malapert Mountain, 180
Mare Tranquillitatis Hole, 185, 190
Mariner 9 spacecraft, 250
Mars: asteroid use, 203, 209, 211–212, 215; atmosphere of, 224, 258, 284, 302; atmospheric pressure on, 251, 254–255, 297–301; bodily changes on, 293–296; comparison with Antarctica, 224, 285, 289–290; comparison with Moon, 173–175, 224–225; crew selection for, 275–278; distance from Earth, 240, 305–306; dust storms on, 250, 262; food security on, 259–265; health risks on, 234–239; homesteading on, 291; human migration to, 279–280; indoor air on, 257–259; landing on, 273–274; NASA plan for, 220, 247–249; possibility of life on, 282–284; profitability, 14, 267–270; radiation danger on, 84, 238–239, 250–251; shelter on, 253–257, 275; technology gaps and, 271–273; temperature on, 249; terraforming on, 296–303; timekeeping on, 285–287; tourism to, 287–289; travel time to, 240–244; water on, 265
Mars 2 spacecraft, 239, 250
Mars 3 spacecraft, 152, 239, 250
Mars500 project, 52–54
Mars City Design, 275
Mars Desert Research Station (MDRS), 226–227
Mars Direct mission concept, 223, 232–234, 241–242, 246, 248
Mars One, 244–245, 265–266
Mars Science Laboratory (MSL), 239, 274
Mars Society, 226–228
mass driver, 164–165, 269
MAVEN spacecraft, 84
McKay, Christopher, 258–259, 262, 299, 303
McMurdo Station, 33, 35
medium-earth orbit (MEO), 127
Mercury, 304, 311–314
Mercury 13 (film), 153n
Merlin rocket engines, 110
metallic hydrogen, 105
meteorites, 34
methane, 225, 229–233, 314, 321, 323–324, 328–329
mice, 77, 87–88, 174, 237–238
MICEHAB, 237–238
microbiome, 86–87, 294–295
microgravity. *See* gravity
Mid-Atlantic Regional Spaceport (MARS), 138
Mimas, 319
mining. *See* Moon: mining on; Mars: profitability
Mir space station, 45, 120, 152, 239
moholes, 302
Mojave Air and Space Port, 136, 138
momentum exchange tether, 140
Moon: Apollo 11 landing, 148, 223; brickmaking on, 177–178; comparison with Antarctica, 31, 160, 181–184, 200; comparison with Mars, 174–175, 224–225; day and night on, 178–179; food security on, 195–198; Kennedy legacy, 6; landscape, 187; lunar dust, 183, 193–195; mining on, 161–172; oxygen extraction on, 185–186; scientific activity on, 172–173;

shelter on, 176–178, 181–185; temperatures on, 178–179; terraforming of, 198–199; tourism to, 186–192; transportation on, 191–192; view of Earth from, 187–188; water and ice abundance on, 162, 180–181
Moon Mineralogy Mapper, 180
Moon Museum, 190–191
moonquakes, 176n
Moon Treaty, 159, 215
Mount Everest, 27–28, 41, 220, 300
MOXIE oxygen generator, 229
Mueller, Tom, 110
Multiple Artificial-gravity Research System (MARS), 77
Mushrooms for Mars, 260
Musk, Elon, 106, 109–113, 245, 281, 297

nanosatellites, 114
NASA: Breakthrough Propulsion Physics (BPP) Project, 341; budget of, 5, 223; Extreme Environment Mission Operations (NEEMO), 47–48; Human Research Program Roadmap, 63–66; Johnson Space Center, 45, 124; Marshall Space Flight Center, 107; 90-Day Study, 230–231; Office of Safety and Mission Assurance, 42; Space Life Sciences Directorate, 96; Space Radiation Laboratory, 88
NASA / Navy Benchmarking Exchange, 42
National Academy of Sciences (NAS), 90
National Defense Education Act of 1958, 151
National Institutes of Health (NIH), 96
National Laboratory, 76
National Space Biomedical Research Institute (NSBRI), 71
National Space Council, 231
Nautilus-X, 77–78
Naval Research Laboratory, 107
Nazis, 99, 150
NEAR spacecraft, 207
Neptune, 305–306, 327, 329–331
NERVA program, 342
New Horizons mission, 305, 307, 331–333
NewSpace, 13, 115
New Swabia, 30
Newton's third law, 131
Nextel fiber, 126
nickel, 203–204, 211
nitrogen, 258–261
nitrogen tetroxide, 104
Nixon, Richard, 5, 9, 223
noise (of rockets and spacecraft), 136–138, 146–147
nuclear fission, 105–106, 342
nuclear fusion, 105–106, 165–168, 334, 342–343
nuclear-powered spacecraft, 103–104, 342–343
nuclear war, 21–22

Obama, Barack, 158, 223, 247
Oberon, 327–329
Occupational Safety and Health Administration (OSHA), 63, 82
Oceanus Procellarum, 185, 190
Olympus Mons, 220, 288–289
O'Neill, Gerard, 117, 127
O'Neill cylinders. *See* orbital cities

Oort cloud, 334–337
Operation Paperclip, 107
Operation Tabarin, 30
Opportunity rover, 250, 271
orbital cities, 5, 117–118, 127, 219, 327
orbital factories, 130
orbital ring, 144–145
orbital velocity, 98–100, 133, 140, 144
Ordovician extinction, 25
Orion spacecraft, 242
orthostatic hypotension, 237
OSIRIS-REx spacecraft, 206
osteoporosis (bone loss), 68–69
'Oumuamua, 23–24
Outer Space Treaty, 31, 159–160, 209–210, 213–215, 280–282
ozone layer, 79–81

Pacific Spaceport Complex–Alaska, 138
Paine, Thomas, 4
Pale Blue Dot (Sagan), 26
Palma, Emilio, 32–33
Parker, Eugene, 86
payload cost, 115
perchlorates, 262–263
Pettit, Don, 97n
Phillips, John, 70
Phobos, 242, 269, 291
phosphorus, 162, 203, 263
PILOT oxygen extractor, 185–186, 228
pintle injector, 110
Planetary Resources, Inc., 208–209, 213
Planetary Society, 340
platinum, 202, 204–205, 210, 268
platinum-group metals, 204
Pluto: as destination, 17, 330–332; distance from Earth, 305–306, 307; gravity on, 327; moons of, 331–332
population, human: disasters and, 21–22, 344; growth of, 20, 346–347
potassium, 162, 263
Project Artemis, 156–157
Project Orion, 103–104, 242–243
Project Timberwind, 342
prophylactic surgery. *See* surgery in space
Proton-M rocket, 113
Proxima Centauri, 305
Psyche (16) asteroid, 204
psychological health risks, 44–45, 50–51, 54–55
pulse-detonation rocket engine, 116

Quayle, Dan, 231

radiation: cosmic, 18, 85–92, 174–176, 224, 250–251, 256; HZE particles, 85, 88; lethal dose, 82–84; solar, 18, 78–85, 174–176, 224, 250–251, 256; units, defined, 81
radiation-induced cognitive decline, 88–89
radio-isotope thermoelectric generators (RTGs), 342
ramjets, 133
rare-earth elements, 162, 168–169
Reagan, Ronald, 170, 222
Rendezvous with Rama (Clarke), 23
RESOLVE oxygen extractor, 228

Return to the Moon (Schmitt), 167
"Return to the Moon" (symposium), 123
Rhea, 320
Robinson, Kim Stanley, 296, 302
rocket equation, 97–100, 104
Rocket Lab, 114–115
Rocket Propellant-1 (RP-1), 104
Roosevelt, Franklin D., 345
Roscosmos, 120
runway, high-altititude, 146
Russia. *See* Soviet accomplishments in space; space race
Ryugu, 207–208, 210–211

SABRE engine, 133
Sadler, Phil, 35–36, 60–61
Sagan, Carl, 26, 311
Salyut space stations, 118–119, 152
Saturn, 305–306, 319–327
Saturn V rocket, 8, 112, 119, 246
Schmitt, Harrison "Jack," 123, 148, 156, 166–167, 174, 176, 189, 194
Schneider, Etienne, 214
Schneider, William, 126
scramjets, 133
SELENE lunar mission, 185
sex in space, 131–132
Shackleton Crater, 161, 179–180
Shackleton Energy Company, 160–161
Shah, Ramille, 273
Shepard, Alan, 4
silicon, 163, 170
silver, 204
Silvera, Isaac, 105
skinsuit, 69
sky crane, 239
skyhooks, 139–142, 269
Skylab, 77, 119–120, 238
Skylon spaceplane, 133–134
Société Européenne des Satellites, 214
sodium, 163
solar electric propulsion, 212
solar energetic particles (SEPs), 80
solar energy, 5, 145, 168, 170, 179, 313
solar flares, 80–81
solar panels. *See* solar energy
solar sails, 336, 340–341
Solar System, scale of, 305–306
solar wind, 78–79, 81, 166, 175
South Pole (Earth), 27, 29, 33–35
Soviet accomplishments in space, 151–152
Soyuz rockets, 113–114, 124, 152
space activity suit (SAS), 252–253
SPACE Act of 2015, 214
Space Adaptation Syndrome (SAS), 235–236
Space Chronicles (Tyson), 102–103
space elevators, 142–144, 269
Space Exploration Initiative (SEI), 230–232
Space Force, 223, 281, 328
space hotels, 128–130
space junk, 144, 158
Space Launch System (SLS), 157–158
spaceplanes, 132–139
Spaceport America, 136–137
spaceports, 136–138
space race: with China, 13, 123, 154–156; with the Soviet Union, 6–8, 119, 150–152
SpaceShipOne, 134
SpaceShipTwo, 135

space shuttle, 4, 9–10, 107–109, 120, 134
Space Shuttle *Challenger,* 231
Space Station *Freedom,* 10, 230–231
Space Station V (fictional), 129
space tourism, 72, 124–125, 130, 186–192, 287–288
SpaceX, 106–116, 123, 125, 186, 245–246
specific impulse, 104
Sphinx-99, 52
Sputnik Crisis, 148–151, 153
Starship (SpaceX spacecraft), 110, 112, 246
"station-keeping," 146
Stone, Bill, 161
Strategic Defense Initiative, 342
Stuster, Jack, 54
submarines, nuclear, 38–44
SUBSAFE, 42
Super Heavy (SpaceX rocket), 110, 112
surgery in space, 92–94
Sylvia asteroid, 202

telomeres, 95
terbium, 162
terrestrial chauvinism, 319, 327
Tesla Roadster, 112–113
Tethys, 319
Tharsis Montes, 288
"third-quarter phenomenon," 51
tholins, 216, 331–332
Thornton, Stan, 120
3-D printing, 130, 176, 218, 273
Tiangong space stations, 121–122
Titan: environment of, 314, 320–325; possibility of life on, 325–327; probe landing on, 14, 321; proposed missions to, 324–325
Titania, 327–329
titanium, 163, 167, 169, 204
Tito, Dennis, 124
Toon, Owen Brian, 21–22
Tranquility Base, 190
TransHub, 126
trans-Neptunian objects, 332–333
Transoceanic Abort Landing sites, 100n
Treaty on Principles Governing the Activities of States in the Exploration and Use of Outer Space. . . . *See* Outer Space Treaty
trinitramide, 104
tritium, 166–167
Triton, 327, 329–330
Tsiolkovsky, Konstantin, 98–99, 142
Tsiolkovsky rocket equation. *See* rocket equation
Twins Study, 94–95
2suit, 131
Tyson, Neil deGrasse, 12, 102–103, 210, 284

ultraviolet (UV) radiation, 78–80
United Launch Alliance (ULA), 109, 111, 114, 161
University of Arizona, 35, 60, 195
uranium, 163, 165, 169–170
Uranus, 298, 305–306, 327–329
US Defense Advanced Research Projects Agency (DARPA), 133
US Department of Defense (DOD), 111, 134, 281

US Government Accountability Office (GAO), 111, 157
US Interstate Highway System, 12, 141

Valles Marineris, 220–221, 288–289
Van Allen radiation belt, 151–152
Vanguard 1, 149–150
VaPER bed-rest study, 48
VASIMR engine, 339–340
Venera spacecraft, 151–152, 308
Venus, 304, 308–311
Vermeulen, Angelo, 50
Verseux, Cyprien, 51
Vesta, 202–203, 208, 216, 218, 339
Viking Landers, 239, 258
Viking Orbiter, 221
Villa Las Estrellas, 32
Virgin Galactic, 114, 135–138
viruses, 93
vision loss, 70
Vonnegut, Kurt, 26
Vostok research station, 34, 179
Voyager 1, 306, 335, 342
V-2 rocket, 99

Walford, Roy, 57–58
warp drive, 341
Webb, James, 6
weightlessness. *See* gravity
Weir, Andy, 236
White, Ed, 3

X3 propulsion engine, 339
X-15 aircraft, 134
X-33 aircraft, 116
X-37 aircraft, 116, 134, 137
X-43 aircraft, 131
X-class flare, 84
xenon, 339
XPRIZE, 154
X-rays, 78–83

Yuegong-1 (Lunar Palace 1), 155, 197
Yutu rover, 154

zero gravity, 67–68
zero-one-infinity rule, 282
Zubrin, Robert: and artificial gravity, 78, 236; on ingenuity and Mars settlement, 269–270; and Mars Direct plan, 223, 232–234, 241–242, 248–249; and Mars Society, 226–227; on skills needed for Mars, 266
Zylon, 140, 269